Pre-K–12 Guidelines for Assessment and Instruction in Statistics Education II (GAISE II)

A Framework for Statistics and Data Science Education

Writing Committee

Anna Bargagliotti (co-chair)
Loyola Marymount University

Christine Franklin (co-chair)
American Statistical Association

Pip Arnold
Karekare Education New Zealand

Rob Gould
University of California Los Angeles

Sheri Johnson
University of Georgia*

Leticia Perez
University of California Los Angeles Center X

Denise A. Spangler
University of Georgia

The Pre-K–12 Guidelines for Assessment and Instruction in Statistics Education II (GAISE II) is an official position of the National Council of Teachers of Mathematics as approved by the NCTM Board of Directors, February 2020.

Endorsed by the American Statistical Association, November 2020.

NCTM | NATIONAL COUNCIL OF TEACHERS OF MATHEMATICS

*Current affiliation -The Mount Vernon School

Library of Congress Cataloging-in-Publication Data

Names: Bargagliotti, Anna, author.

Title: Pre-K–12 guidelines for assessment and instruction in statistics
education II (GAISE II) / writing committee, Anna Bargagliotti
(co-chair), Loyola Marymount University, Christine Franklin (co-chair),
American Statistical Association, Pip Arnold, Karekare Education New
Zealand, Rob Gould, University of California Los Angeles, Sheri Johnson,
University of Georgia, Leticia Perez, University of California Los
Angeles, Denise A. Spangler, University of Georgia.

Other titles: Guidelines for assessment and instruction in statistics
education (GAISE) report

Description: Second edition. | Alexandria, VA : American Statistical
Association, 2020. | Includes bibliographical references. | Summary:
"This document lays out a curriculum framework for Pre-K–12 educational
programs that is designed to help students achieve data literacy and
become statistically literate. The framework and subsequent sections in
this book recommend curriculum and implementation strategies covering
Pre-K–12 statistics education"-- Provided by publisher.

Identifiers: LCCN 2020006082 | ISBN 9781734223514 (paperback)

Subjects: LCSH: Statistics--Study and teaching (Early
childhood)--Standards. | Statistics--Study and teaching
(Elementary)--Standards. | Statistics--Study and teaching
(Secondary)--Standards.

Classification: LCC QA276.18 .G85 2020 | DDC 519.5071/2--dc23

LC record available at https://lccn.loc.gov/2020006082

10 9 8 7 6 5 4 3 2 1

978-1-7342235-1-4

Contents

Acknowledgments v

Preface 1

Introduction 5

Framework 13

Level A 21

   Introduction 22

   Essentials for Each Component 22

   Example 1: Choosing the Band for the End of the Year Party –
                Conducting a Survey and Summarizing Data 24

   Example 2: Family Size - Mean as Equal/Fair Share and Variability as Number of Steps 28

   Example 3: What do Ladybugs Look Like – Collecting, Summarizing, and Comparing Data 31

   Example 4: Growing Beans – A Simple Comparative Experiment 35

   Example 5: Growing Beans (Continued) – Time Series 37

   Example 6: CensusAtSchool – Using Secondary Data and Looking at Association 38

   Summary of Level A 41

Level B 43

   Introduction 44

   Essentials for Each Component 44

   Example 1: Level A Revisited: Choosing the Music for the School Dance –
                Multivariable and Larger Groups 45

   Example 2: Choosing Music for the School Dance (Continued) – Comparing Groups 49

   Example 3: Choosing Music for the School Dance (Continued) –
                Connecting Two Categorical Variables 51

   Example 4: Darwin's Finches – Comparing a Quantitative Variable Across Groups 53

   Example 5: Darwin's Finches (Continued) – Separation Versus Overlap 57

   Example 6: Darwin's Finches (Continued) – Measuring the Strength of Association
                Between Two Quantitative Variables 59

   Example 7: Darwin's Finches (Continued) – Time Series 61

   Example 8: Dollar Street – Pictures as Data 62

   Example 9: Memory and Music – Comparative Experiment 67

   Summary of Level B 68

Level C **71**

Introduction 72

The Role of Technology 73

The Role of Probability in Statistics Essentials for Each Component 73

Example 1: Darwin's Finches from Level B (Continued) – MAD to Standard Deviation 77

Example 2: Level A and B Revisited: Choosing Music for the School Dance – Generalizing Findings 79

Example 3: Choosing Music for the School Dance (Continued) – Inference About Association 83

Example 4: Effects of Light on the Growth of Radish Seedlings – Experiments 84

Example 5: Considering Measurements When Designing Clothing – Linear Regression 89

Example 6: Napping and Heart Attacks – Inferring Association from an Observational Study 92

Example 7: Working-age Population – Working with Secondary Data 94

Example 8: Classifying Lizards – Predicting a Categorical Variable 97

Summary of Level C 102

Assessment **105**

National and International Standardized Assessments 105

Sources of Quality Items for Educators 105

Level A Assessment Examples 106

Level B Assessment Examples 108

Level C Assessment Examples 111

References **115**

GAISE II Acknowledgements

A special thank you to Christine Franklin, Gary Kader, Denise Mewborn (Spangler) Jerry Moreno, Roxy Peck, Mike Perry, and Richard Scheaffer for their leadership and vision in the first seminal GAISE I document.

The authors extend a sincere thank you to the ASA/NCTM Joint Committee for funding the writing and production process of GAISE II. A special thank you to Donna LaLonde and Rebecca Nichols from ASA and Dave Barnes and Jeff Shih from NCTM for their support throughout the process.

Thank you to Brenna Bastian for her beautiful artwork on the cover of the report and the Level A Ladybugs Example. We also appreciate the design and layout work of Shirley E.M. Raybuck and Valerie Nirala.

Lastly, the authors would like to express their heartfelt appreciation to the twenty-two people who closely reviewed this GAISE II document and provided valuable feedback:

Gail Burrill, Rosemarie Callingham, Catherine Case , Michelle Dalrymple, Neville Davies, Ed Dickey, David Fluharty, Gary Kader, Donna LaLonde, Jerry Moreno, Rebecca Nichols, Regina Nuzzo, Roxy Peck, Jamis Perrett , Maxine Pfannkuch, Katherine Respress, Richard Schaeffer, Neil Sheldon, Josh Tabor, Dan Teague, Doug Tyson, and Jane Watson.

GAISE II Preface

In 2020 as the *Pre-K–12 Guidelines for Assessment and Instruction in Statistics Education II: A Framework for Statistics and Data Science Education report (GAISE II)* is published, never have data and statistical literacy been more important. The public is being called upon to synthesize information from many global issues, including the COVID-19 global pandemic, a changing planet with extreme weather conditions, economic upturns and downturns, and important social issues such as the Black Lives Matter movement. Data are encountered through visualizations (sometimes interactive and sometimes not), reports from scientific studies (such as medical studies), journalists' articles and websites.

The demands for statistical literacy have never been greater. Statistically literate high-school graduate need to be able to evaluate the conclusions and legitimacy of reported results as well as formulate their own analyses. Steve Levitt, co-author of *Freakonomics*, addressed the need for statistical and data literacy with this quote from an October 2, 2019 podcast:

> *I believe that we owe it to our children to prepare them for the world that they will encounter—a world driven by data. Basic data fluency is a requirement not just for most good jobs, but also for navigating life more generally, whether it is in terms of financial literacy, making good choices about our own health, or knowing who and what to believe.*

Driven by the digital revolution, data are now readily accessible to statistical methods and technological tools so that students can gain insights and make recommendations to manage pressing world issues. Data can be extremely valuable, but only if they are used judiciously and in a proper context.

Today, many sectors of the economy and most jobs rely on data skills. Good data sense is needed to easily read the news and to participate in society as a well-informed member. Because of this, it is essential that all students leave secondary school prepared to live and work in a data-driven world. The Pre-K–12 GAISE II report presents a set of recommendations for school-level statistical literacy.

Overview and Goals of GAISE II

Guidelines for Assessment and Instruction in Statistics Education: A Pre-K–12 Curriculum Framework (hereafter referred to as Pre-K–12 GAISE I) was first released in 2005 with slight revisions in 2007 (https://www.amstat.org/asa/files/pdfs/GAISE/GAISEPreK-12_Full.pdf) along with the *Guidelines for Assessment and Instruction in Statistics Education College Report.* The *GAISE College Report* outlined recommendations for the post-secondary introductory statistics course and was updated in 2016 (https://www.amstat.org/asa/files/pdfs/GAISE/GaiseCollege_Full.pdf).

The Pre-K–12 GAISE I report was written to enhance the statistics standards in the National Council for Teachers of Mathematics (NCTM) 2000 *Principles and Standards of School Mathematics* and as a follow-up document to the Conference Board of Mathematical Sciences (CBMS) *Mathematical Education for Teachers (MET)* document. The Pre-K–12 GAISE I was a seminal and visionary document that championed the necessity of data and statistical literacy starting in the early school grades. It provided a framework of recommendations for developing students' foundational skills in statistical reasoning in three levels across the school years, described as levels A, B, and C. These levels are maintained in GAISE II and are roughly equivalent to elementary, middle, and high school. The progression through the sequential levels in the Pre-K–12 GAISE (both I and II) is intended for any individual who is striving to achieve statistical literacy, regardless of age.

Since its initial publication, GAISE I has significantly impacted the inclusion of statistics standards at the state and national level in the United States and across the world. The report has been used internationally as a reference point for statistics education at the school level. There is a Spanish translation of GAISE I (https://www.amstat.org/asa/files/pdfs/GAISE/Spanish.pdf). At the time of this writing, Google scholar shows over 790 citations for GAISE I in scholarly works. The report has been referenced in numerous National Science Foundation grant projects and other professional STEM educational organizations' reports. GAISE I also influenced the development of state standards across the United States and the writing of the ASA *Statistical Education of Teachers* (SET) document (https://www.amstat.org/asa/files/pdfs/EDU-SET.pdf) that makes recommendations for the preparation of school level teachers in statistics.

GAISE I primarily focused on traditional data types of quantitative and categorical variables and on study designs using small data sets of samples from a population. Fifteen years later, data types have expanded beyond being classified as quantitative and categorical thus necessitating the acquisition of different and often state-of-the-art statistical skills. Today, for example, data include text posted on social media or highly structured (or unstructured) collections of pictures, sounds, or videos. Data are immense and readily available. Data are multidimensional. Data representations and visualizations are also often multidimensional and interactive displaying many variables simultaneously.

GAISE II incorporates the new skills needed for making sense of data today while maintaining the spirit of GAISE I. GAISE II highlights:

1. The importance of asking questions throughout the statistical problem-solving process (formulating a statistical investigative question, collecting or considering data, analyzing data, and interpreting results), and how this process remains at the forefront of statistical reasoning for all studies involving data

2. The consideration of different data and variable types, the importance of carefully planning how to collect data or how to consider data to help answer statistical investigative questions, and the process of collecting, cleaning, interrogating, and analyzing the data

3. The inclusion of multivariate thinking throughout all Pre-K–12 educational levels

4. The role of probabilistic thinking in quantifying randomness throughout all levels

5. The recognition that modern statistical practice is intertwined with technology, and the importance of incorporating technology as feasible

6. The enhanced importance of clearly and accurately communicating statistical information

7. The role of assessment at the school level, especially items that measure conceptual understanding and require statistical reasoning involving the statistical problem-solving process

A Future Driven by Data

As stated in GAISE II,

> *Data are used to tell a story. Statisticians see the world through data – data serve as models of reality. Statistical thinking and the statistical problem-solving process are foundational to exploring all data.*

GAISE II presents a vision where every individual is confident in reasoning statistically, making sense of data, and knowing how and when to bring a healthy skepticism to information gleaned from data. Presented here is a framework of essential concepts and 22 examples across the three levels of skills development. This framework supports all students as they learn to appreciate the vital role of statistical reasoning and data science and acquire the essential life skill of data literacy.

We, the writers, are appreciative of the opportunity, support, and endorsement from the Board of Directors of both the American Statistical Association (ASA) and the National Council for the Teachers of Mathematics (NCTM) for the enhancement of the Pre-K–12 GAISE I document. Our hope is that the Pre-K–12 GAISE II document will enrich your work in fostering the ultimate goal: statistical literacy for all.

Anna Bargagliotti (co-chair)
Christine Franklin (co-chair)
Pip Arnold
Rob Gould
Sheri Johnson
Leticia Perez
Denise A. Spangler

Introduction

Our lives are heavily influenced by data. Every person should be able to use sound statistical reasoning to intelligently make evidence-based decisions. We need statistical literacy to succeed at work, stay informed about current events, and be prepared for a healthy, happy, and productive life.

Each morning, the newspaper and other media confront us with statistical information on topics ranging from the economy to education, from movies to sports, from food to medicine, and from public opinion to social behavior. Such information guides decisions in our personal lives and enables us to meet our responsibilities as members of a community and society. Statistical literacy is a requirement for navigating today's world.

Data can also provide support in our personal lives. Wearing a fitness device allows us to track steps, heart rate, and other fitness statistics in a way that motivates healthier lifestyles. If we consider moving to another community, we could make decisions based on statistics about the cost of living or local climate.

Statistically literate individuals may have the opportunity to advance in their careers and obtain more rewarding and challenging positions. Business leaders may be presented with quantitative information on budgets, supplies, manufacturing specifications, market demands, sales forecasts, or workloads. Teachers may be confronted with educational statistics concerning student performance or their own accountability. Medical scientists must understand the statistical results of experiments used for testing the effectiveness and safety of medical treatments. Law enforcement professionals depend on crime statistics.

Statistical literacy involves having a healthy dose of skepticism about findings based upon data. A statistically literate high school graduate will be able to evaluate conclusions from data and judge the legitimacy of reported results. The Business Higher Education Forum (BHEF) report *The New Foundational Skills of the Digital Economy* agrees and states (www.bhef.com/sites/default/files/BHEF_2018_New_Foundational_Skills.pdf):

> *At some moment in the future, many of the high levels of skill that currently seem confined to the upper reaches of the digital economy, or to larger, more complex organizations, will become the norm among jobseekers, incumbent employees, and workplaces. (BHEF 2018 publication, p. 18)*

The ultimate goal: statistical literacy for all.

Sound statistical skills take time to develop. They cannot be honed to the level needed in the modern world through one high school course. The best way to help people attain statistical literacy is to begin their statistics education in the elementary grades and keep strengthening and expanding students' statistical skills throughout middle school and high school years.

This document lays out a curriculum framework for Pre-K–12 educational programs that is designed to help students achieve data literacy and become statistically literate. The framework and the subsequent sections in this report recommend curriculum and implementation strategies covering Pre-K–12 statistics education.

While the document provides a Pre-K–12 education framework, the pathway to statistical literacy presented in this document is appropriate for any age bracket. In fact, the document's framework is meant for any individuals working towards statistical literacy. Throughout the report these individuals are referred to as simply "students."

Why Guidelines for Assessment and Instruction in Statistics Education (GAISE) II? Overview and Goals of GAISE II

The Guidelines for Assessment and Instruction in Statistics Education: A Pre-K–12 Curriculum Framework (GAISE I) was first released in 2005 with a slight revision in 2007 (see Figure 1). This was a groundbreaking document. It made the case for statistical literacy at the school-level and explained how mathematical and statistical thinking are related yet distinct. Mathematical skills are necessary, and statistical and mathematical thinking must work in concert when analyzing data.

Since GAISE I was published, traditional conceptions of data have changed. Data are no longer simply numbers in context, classified as quantitative or categorical, typically stored in static spreadsheets. Today, data can also be dynamic, complex, highly structured (or unstructured) collections of pictures or sounds. Data sets are vast and readily available.

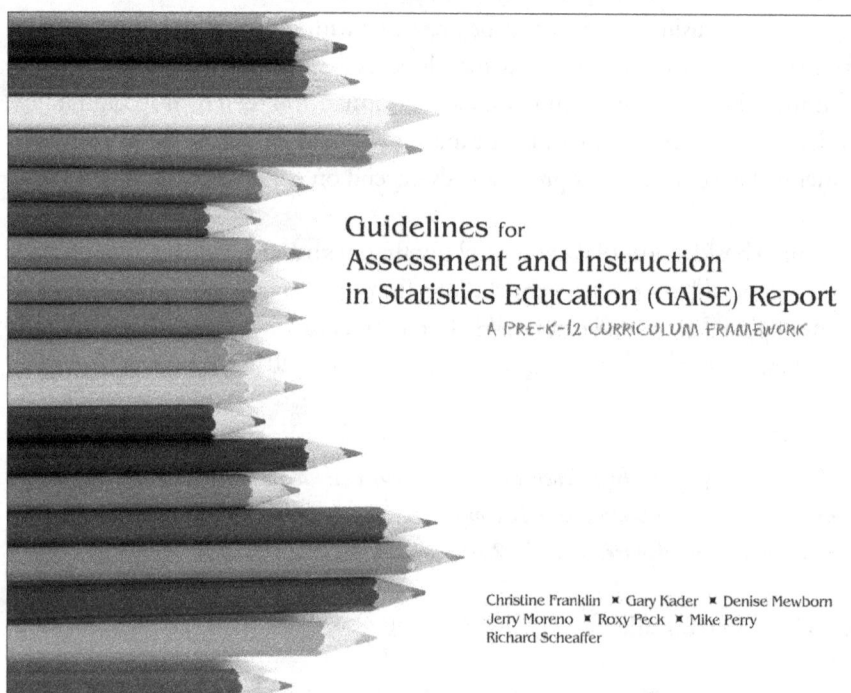

Guidelines for
Assessment and Instruction
in Statistics Education (GAISE) Report
A PRE-K-12 CURRICULUM FRAMEWORK

Christine Franklin ✕ Gary Kader ✕ Denise Mewborn
Jerry Moreno ✕ Roxy Peck ✕ Mike Perry
Richard Scheaffer

Figure 1: GAISE I was first released in 2005.

GAISE II includes examples that deal with non-traditional data and multivariable data throughout the entire Pre-K–12 curriculum. Students must develop statistical literacy to make sense of the immense amount of data that surround them on a daily basis. Much of these data are generated by students themselves through social media, global positioning system (GPS) devices, and so on. Students should become aware of how these data are stored and how they are utilized by the organizations collecting them. They should also understand why our society needs security measures and policies to prevent the mishandling and unethical use of these data.

Students need to begin at an early age to become data-savvy, whether working with small data sets or large, messy data sets, with traditional data or non-traditional data such as text or images. Most future jobs will require some knowledge of statistics and data analytics.

Being able to reason statistically is essential in all disciplines of study and work. Many disciplines such as the sciences now include statistics in their standards (e.g., Next Generation Science Standards (NGSS) Lead States, 2013). Several of the new examples in GAISE II use science data sets and touch on topics discussed in the Next Generation Science Standards (NGSS) to illustrate how statistical reasoning is an integral component of scientific investigations.

While progress has been made over the past 50 years in preparing students, there is still much work necessary for the future as the discipline of statistics continues to evolve. Some important documents since the publication of GAISE I promoting statistical literacy at the Pre-K–12 school-level are The American Statistical Association (ASA)'s *Statistical Education of Teachers (SET)* published in 2015 and National Council of Teachers of Mathematics (NCTM)'s *Catalyzing Change: Initiating Critical Conversations* books for elementary, middle (2020), and high school (2018). Those interested in learning about the history of statistics in the Pre-K–12 setting may refer to Chapter 9 in SET.

The spirit of the original GAISE I report remains. The statistical problem-solving process defined in GAISE I remains the foundation and core of statistical reasoning and making sense of data. The statistical problem-solving process is defined as a four-step process of (1) formulating a statistical investigative question, (2) collecting or considering data, (3) analyzing data, and (4) interpreting results. The three levels of A, B, and C (loosely meant to match elementary, middle, and high school) are also consistent across the GAISE I and GAISE II reports.

Spurred by the overabundance of data available in today's world, the statistical problem-solving process not only remains important, it becomes even more critical for drawing conclusions from data. This includes recognizing misleading graphical representations and limitations of data sets, no matter how large, being used to answer statistical investigative questions. For examples of misuses in statistics and limitations of data sets see the following: The New York Times Magazine's *When the Revolution Came for Amy Cuddy* (www.nytimes.com/2017/10/18/magazine/when-the-revolution-came-for-amy-cuddy.html) and Kuiper's *Incorporating Research Experience into an Early Undergraduate Statistics Course* (http://iase-web.org/documents/papers/icots8/ICOTS8_4G1_KUIPER.pdf?1402524970).

Statistically sound studies should be reproducible. Repeated studies with similar samples should yield similar conclusions, and different statistical methods re-applied to the same data should give consistent results.

As emphasized in the original GAISE, "statistics is a methodological discipline. It exists not for itself, but rather to offer other fields of study a coherent set of ideas and tools for dealing with data. The need for such a discipline arises from the *omnipresence of variability*" (Cobb & Moore, 1997, p. 801). Statistical thinking, in large part, must deal with the omnipresence of variability in data (e.g., variability within a group, variability between groups, sample-to-sample variability in a statistic). Statistical problem solving and decision making depend on understanding, explaining, and quantifying variability in the data within the given context. "Statistics requires a different *kind* of thinking, because data are not just numbers, they are numbers with a context" (Cobb & Moore, 1997, p. 801). "In mathematics, context obscures structure. In data analysis, context provides meaning" (Ibid, p. 803).

To highlight the importance of statistical literacy today, consider the following graph in Figure 2 from *The New York Times's* What's Going On in This Graph? from April 9, 2018.

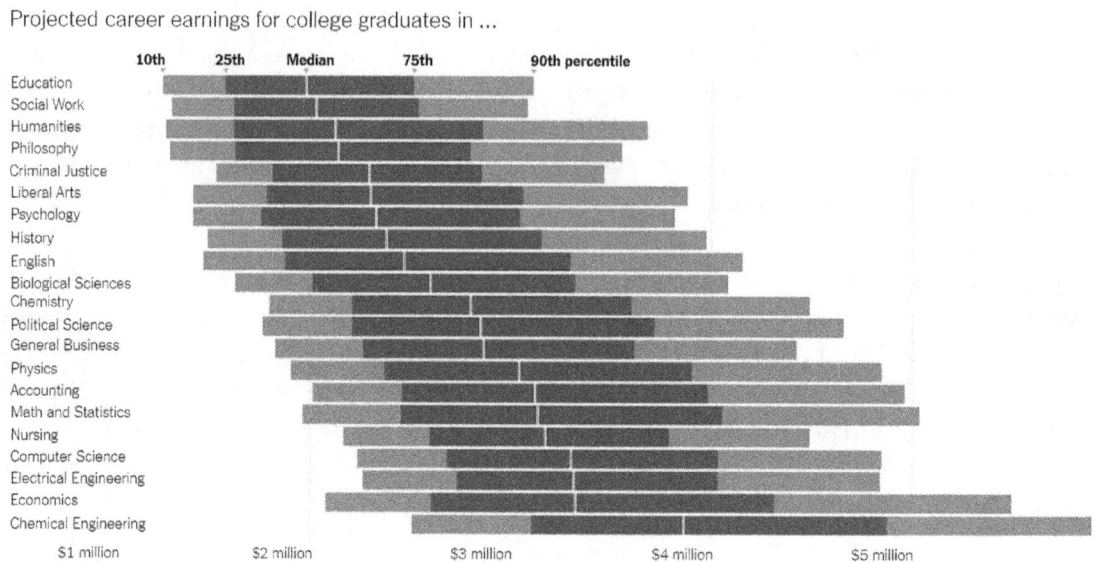

Projected career earnings for college graduates in ...

Figure 2: : Distributions of projected earnings for different professions.
Source: *The New York Times's* What's Going On in This Graph? April 9, 2018
"The Lifetime Earnings Premia of Different Majors," 2014 (updated: 2017), by Douglas A. Webber

This graphic displays the distributions of projected earnings for different professions. The distributions are represented with bars where the bars are divided into segments by the 10th, 25th, 50th, 75th, and 90th percentiles (measures of positions for each distribution). Both the context and variability of the different distributions are important when interpreting this graphic. For example, we can notice that education is one of the lower-paying professions compared to chemical engineering, but there is also more variability for the salaries of chemical engineers (note the length of the segments within the bars). Context is important, too. We notice that this graphical display provides little information about how the data were obtained. It is important to question how the data were obtained – are they primary data collected by researchers or are they secondary data made available to investigators?

It is these features—the focus on variability in data, the importance of context associated with the data, and the questioning of data—that sets statistics apart from other mathematical sciences and makes it particularly relevant for all fields of study.

It is critical that statisticians—or anyone who uses data—be more than just data crunchers. They should be data problem solvers who interrogate the data and utilize questioning throughout the statistical problem-solving process to make decisions with confidence, understanding that the art of communication with data is essential.

This report, therefore, enhances and updates the GAISE I report of 2005 and 2007 to adjust for the remarkable evolution within the statistical field over the past 15 years. These enhancements include an emphasis on:

• Questioning throughout the statistical problem-solving process

• Different data and variable types

- Multivariable thinking throughout Levels A, B, and C

- Probabilistic thinking throughout Levels A, B, and C

- The role of technology in statistics and how it develops throughout the Levels

- Assessment items that measure statistical reasoning

Questioning in Statistics

Regardless of the kind of data with which we're working — whether small samples from a population, experimental data . . . addressing each component of the process is essential.

The statistical problem-solving process typically starts with a statistical investigative question, followed by a study designed to collect data that aligns with answering the question. Analysis of the data is also guided by questioning. Constant questioning and interrogation of the data throughout the statistical problem-solving process can lead to the posing of new statistical investigative questions.

Often when considering secondary data, the data need to first be interrogated – how were measurements made, what type of data were selected, what is the meaning of the data, and what was the study design to collect the data. Once a better understanding of the data has been gained, then one can judge whether the data set is appropriate for exploring the original statistical investigative question or one can pose statistical investigative questions that can be explored with the secondary data set.

GAISE II models the use of questioning in statistics in all its examples. For a more detailed discussion on the role of questioning and the different types of statistical questions used in the statistical problem-solving process, see Arnold & Franklin (2020).

Different Data and Variable Types

Traditional variables types are classified as categorical or quantitative (numerical). Categorical variables can be further classified as either nominal or ordinal (ranking). Likewise, quantitative variables can be described by whether they are measured on discrete or continuous scales. Counts, such as the number of pets a student has, are examples of a discrete quantitative variable. Measurements, such as the length of a lizard's tail, are examples of a continuous quantitative variable. It is essential that students become comfortable with analyzing these traditional variable types. They should learn how to explore and describe the characteristics of a data distribution for categorical and quantitative variables.

In our modern-day world, variables can also be pictures, sound, video, or words. Students need to be able to identify raw data of these non-traditional variable types, understand how variable transformations can produce different representations of the same data, and organize these data appropriately. Students will need to ask how the data are produced – are they primary (data that are collected first hand) or secondary (data that are available).

More data are being captured and secondary data sources abound. Such data are often available but not ready for analysis, and modern students must gain skills in manipulating and restructuring data, transforming provided variables into new variables, and querying the origins and suitability of data for the purpose at hand (e.g., asking whether social media posts can be generalized to the larger population). Examples throughout this document will illustrate how to work with different variable types.

Multivariable Thinking in Statistics

Multivariable thinking should begin at an early age as it is natural for students to question and reason with more than two variables at a time, especially exploring associations between variables. Young students may notice multiple features of an observational unit, such as themselves and people in their class, and record these data. All students, even those at the youngest ages, should be encouraged to draw comparisons between groups. As students advance in their statistical learning, they develop the statistical tools to formalize comparisons and associations.

Probability in Statistics

Probability is about quantifying randomness. It is the foundation for how we make predictions in statistics. Beginning at an early age, students can use probability informally to predict how likely or unlikely particular events may be. They can also consider informal predictions beyond the scope of the data that they have analyzed.

Probability is an essential tool in statistics. Probability is also important in mathematics which employs different approaches and different reasoning than that used in statistics. Two problems and the nature of the solutions will illustrate the difference.

Problem 1: Assume a die is "fair."
> *Question:* If a die is rolled 10 times, how many times will we see an even number on the top face?

Problem 2: You pick up a die.
> *Question:* Is this a fair die? That is, does each face have an equal chance of appearing?

Problem 1 is a mathematical probability problem. Problem 2 is a statistics problem that can use the mathematical probability model determined in Problem 1 as a tool to seek a solution.

The answer to neither question is deterministic. Dice rolling produces random outcomes, which suggests that the answer is probabilistic. The solution to Problem 1 starts with the assumption that the die is fair. It then proceeds to logically deduce the numerical probabilities for each possible count of even numbers resulting from 10 rolls. The possible counts are 0, 1, …, 10.

The solution to Problem 2 starts with an unfamiliar die; we do not know if it is fair or biased. The search for an answer is experimental: roll the die, see what happens, and examine the resulting data to see whether they look as if they came from a fair die or a biased die. One possible approach to making this judgment would be the following: Roll the die 10 times and record the number of even numbers that are rolled. Repeat this process of rolling the die 100 times. Compile the frequencies of even-numbered rolls for each of these 100 trials (e.g., 5, 3, 6, …). Compare these results to the frequencies predicted by the mathematical model for a fair die in Problem 1. If the empirical frequencies from the experiment are quite dissimilar from those predicted by the mathematical model for a fair die and the observed frequencies are not likely to be due to chance variability, then we can conclude that the die is not fair.

In Problem 1 we form our answer from logical deductions. In Problem 2, we form our answer by observing experimental results. For similar ideas exploring basic notions of probability, see Gelman & Nolan (2002).

Probability calculations can often be approximated through simulation, both by hand and with technology. Simulations help students get a conceptual understanding of complex probability calculations without relying on mathematical computation. Level C contains a more enhanced discussion of the role of probability for statistical reasoning.

Probability is also used in statistics through randomization – random sampling and random assignment. Samples can be collected at random and experiments can be designed by randomly assigning individuals to different treatments. Randomization minimizes bias in selections and assignments. It also leads to random chance in outcomes that can be described with probability models.

Technology in Statistics

The teaching of statistics has been greatly enhanced, moving from teaching with no technology to teaching with integrated technology. The field has evolved from using programming languages in the 1980s to hand-held statistical calculators in the 1990s to online statistical calculators, powerful statistical software packages, and amazing data visualization tools. Simulation is now as easy as accessing a public applet where point-and-click options provide the ability to perform thousands of trials. Computer labs are not necessary – just internet access. Moving to web-based technology allows more access to data visualization, exploration of data, and simulation. However, access to technology varies across school districts. Not all classrooms are equipped with internet access or technology hardware and software. Modern statistical practice is intertwined with technology; thus, it is recommended that technology be embraced to the greatest extent possible within a given circumstance.

GAISE II illustrates how to incorporate appropriate uses of technology into statistical activities in Pre-K–12. Level C also contains a more detailed discussion of how practicing statisticians use technology.

Assessment in Statistics

Regardless of the type of assessment used, assessment items should measure conceptual understanding. They must require students to use statistical reasoning with context and variability at all stages of the statistical problem-solving process. GAISE II provides examples from national and international projects that model this kind of assessment.

Assessments should also align the technology used for statistical computation with the technology used for teaching. For example, if statistical software is used for teaching, the same software should be used for assessments as opposed to having a mismatch, and for example, only allowing a calculator on the assessment. Furthermore, technology for the administration of assessments, both formative and summative should be considered.

The Future with Data

Today, the art and science of working with data appears under many names, including statistics, data science, informatics, and data analytics. These all combine the skills of statistics, mathematics, and computer science. With the abundance of data that are collected daily through the internet and other media, machine learning, deep learning, and artificial intelligence are growing areas where large data sets are used and algorithms are developed to make predictions. For example, algorithms for predicting consumer behavior are used in marketing. With these areas of study, it is critical that the statistical problem-solving process continually be utilized to interrogate the data. Without this interrogation, biases and misuses might emerge. See for example ProPublica's *Machine Bias* which reveals inequities in the criminal justice system (www.propublica.org/article/machine-bias-risk-assessments-in-criminal-sentencing)

Data in both the public and private sectors surround and shape us in our professional and personal lives. Data are a means of communication, community building, and discovery. Data are used to tell a story. Statisticians see the world through data – data serve as models of reality. Statistical thinking and the statistical problem-solving process emphasized in GAISE II are foundational to exploring all data.

Framework

The conceptual structure for statistics education is provided in the two-dimensional framework model shown in Table 1. One dimension is defined by the statistical problem-solving process components that can be used to advance statistical literacy. The second dimension is composed of three developmental levels.

Statistical Problem-Solving Process

The purpose of the statistical problem-solving process (see Figure 3) is to collect and analyze data to answer statistical investigative questions.

This investigative process involves four components, each of which involve exploring and addressing variability:

I.   Formulate Statistical Investigative Questions

II. Collect/Consider the Data

III. Analyze the Data

IV. Interpret the Results

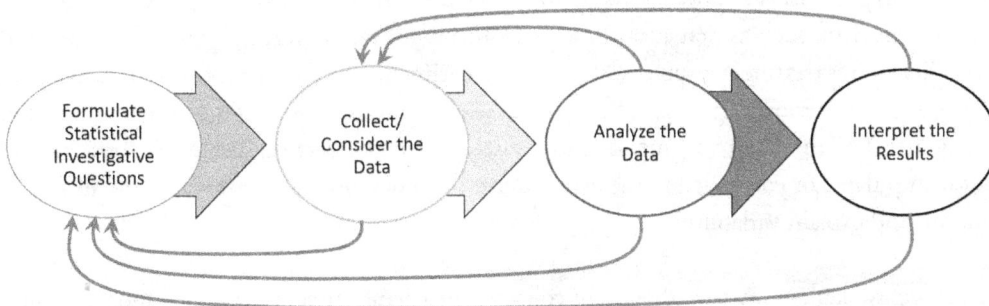

Figure 3: Statistical problem-solving process

I. Formulate Statistical Investigative Questions

Anticipating Variability – Beginning the Process
Formulating statistical investigative questions that anticipate variability leads to productive investigations. For example, the following are all statistical investigative questions that anticipate variability and could lead to a rich data collection process and subsequent analysis of the data:

- *How fast will my plant grow?*

- *Do plants exposed to more sunlight grow faster?*

- *How does sunlight affect the growth of a plant?*

In contrast, the question *How tall is the plant?* is answered with a single height; it is therefore not a statistical investigative question. *How tall is the plant* is a question we ask to collect data. Many other data collection questions could be asked to help collect the necessary data to answer the statistical investigative question: *Do plants exposed to more sunlight grow faster?* The fact that there will be differing heights for the different exposures of sunlight implies that we anticipate an answer based on measurements of plant heights that vary.

While statistical investigative questions begin worthwhile studies, the use of questioning is prominent throughout all four components of the statistical problem-solving process. Such uses of questioning will be illustrated throughout the examples at the different levels.

In addition to anticipating variability, there are other features of a statistical investigative question that are important. The variable(s) of interest must be clear; the group or population that the question is focused on must be clear; the intent of the question should be clear – is the question requiring a description of the data, is the question comparing a variable across two or more groups, is the question looking at an association between two variables; the question should be about the whole group (anticipating variability) and not about an individual (giving a deterministic answer); the question should be answered through data collection (primary data) or with the data in hand (secondary data); and the question should be purposeful.

II. Collect/Consider Data
Acknowledging Variability—Designing for Differences

Data collection designs must acknowledge variability in data. Some study methods are used to reduce and detect variability in data, such as Statistical Process Control and random sampling. Others are used to induce variability to test treatments, such as Design of Experiments. In the latter approach, experimental designs are chosen to acknowledge the differences between groups subjected to different treatments. Random assignment to the groups is intended to reduce differences between the groups due to factors that are not manipulated or controlled in the experiment. In all designs, a main statistical focus is to look for, account for, and explain variability.

After the data are available—whether they were collected first-hand or acquired from another source— they need to be interrogated. For example, questions about how the variables differ by type, the possible outcomes of each of the variables, and how the data were collected are necessary to clarify whether the data are useful for answering the statistical investigative question. The data collection design impacts the scope of generalizability and the possible limitations in analysis and interpretation.

III. Analyze the Data
Accounting of Variability—Using Distributions

When we analyze data, we seek to understand its variability. Reasoning about distributions is key to accounting for and describing variability at all developmental levels. Graphical displays and numerical summaries are used to explore, describe, and compare variability in distributions.

For example, the batting averages of the American League baseball teams and the batting averages of the National League baseball teams for a particular year can be displayed in two comparative dotplots and boxplots. These graphs show the variability of each league's distribution of batting averages. We can take into account variability by describing the overlap and the separation of the distributions of the two leagues.

Another example of taking variability into account is the margin of error in public opinion polling. When the results of an election poll state that "42% of those polled support a particular candidate with a margin of error of +/- 3 percentage points at the 95% confidence level," the focus of the margin of error is to account for sampling variability.

IV. Interpret the Results
Allowing for Variability—Looking beyond the Data

Statistical interpretations are made in the presence of variability and must take variability into account. For example, we should interpret the result of an election poll as an estimate that may vary from sample to sample of voters being polled. When interpreting the results of a randomized comparative medical experiment, we must remember there are two important sources of variability: randomization to treatment group, and variability from individual to individual. When we generalize the results and look beyond the study data collected, we must take into account these sources of variability.

Three Developmental Levels: A, B, and C

Experienced statisticians understand the role of variability in the statistical problem-solving process. When they formulate their first question, they anticipate the data collection, the nature of the analysis, and the possible interpretations—all of which involve possible sources of variability. In the end, mature practitioners reflect upon all aspects of data collection and analysis as well as the question itself when interpreting results. Likewise, they link data collection and analysis to each other as well as to the other components in the statistical problem-solving process.

Beginning students cannot be expected to make all of these linkages. They require years of experience and training to develop more mature reasoning. Much like mathematics education, statistics education should be viewed as a developmental process.

As in GAISE I, to meet the goals of statistical literacy, this report provides a framework for statistical education within Pre-K–12 settings over three Levels, A, B, and C. Students at very young ages innately have notions of variability and probability. Related research is summarized in *Statistics in Early Childhood and Primary Education* (Leavy, Meletiou-Mavrotheris & Paparistodemou, 2018). Level A capitalizes on these understandings by more formally introducing students to the statistical problem-solving process. Level B continues to build the statistical toolbox. By the time students reach Level C, the student can be provided with ambitious learning goals towards the development of statistical literacy in Pre-K–12 education. Level C sets lofty goals for students finishing Pre-K–12 education in today's data driven society. This sets the stage for students to mature past these levels to further develop more complex statistical investigative questions and analysis techniques while working with ever-evolving data types.

Although these three levels may parallel grade levels, they are based on development in statistical literacy, not age. There is <u>no</u> attempt to tie these levels to specific grade levels. Thus, a middle school student who has had no prior experience with statistics will need to begin with Level A concepts and activities before moving to Level B. This prerequisite holds for a secondary student as well. If a student has not had Level A and B experiences prior to high school, then it is not appropriate for that student to begin with Level C expectations. Investigations and scenarios are more teacher-driven at Level A but become more student-driven at Levels B and C.

The Framework Table

As was structured in the framework table from GAISE I, each of the four stages or process components is described as it develops across levels. It is understood that work at Level B assumes and develops further the concepts from Level A; likewise, Level C assumes and uses concepts from the lower levels. The essentials from GAISE I are similar in GAISE II but enhanced with more specifics and some additional essentials to account for the evolution of the statistical field since GAISE I.

Reading down a column will describe a complete problem investigation for a particular level.

Table 1: The Framework

Process Component	Level A	Level B	Level C
I. Formulate Statistical Investigative Questions	Understand when a statistical investigation is appropriate	Recognize that statistical investigative questions can be used to articulate research topics and that multiple statistical investigative questions can be asked about any research topic	Formulate multivariable statistical investigative questions and determine how data can be collected and analyzed to provide an answer
	Pose statistical investigative questions of interest to students where the context is such that students can collect or have access to all required data	Understand that statistical investigative questions take into account context as well as variability present in data	Pose summary, comparative, and association statistical investigative questions for surveys, observational studies, and experiments using primary or secondary data
	Pose summary (or descriptive) statistical investigative questions about one variable regarding small, well-defined groups (e.g., subset of a classroom, classroom, school, town) and extend these to include comparison and association statistical investigative questions between variables	Pose summary, comparative, and association statistical investigative questions about a broader population using samples taken from the population	Pose inferential statistical investigative questions regarding causality and prediction
	Experience different types of questions in statistics: those used to frame an investigation, those used to collect data, and those used to guide analysis and interpretation	Pose statistical investigative questions that require looking at a variable over time	
		Understand that there are different types of questions in statistics: those used to frame an investigation, those used to collect data, and those used to guide analysis and interpretation	
		Pose statistical investigative questions for data collected from online sources and websites, smartphones, fitness devices, sensors, and other modern devices	

Table 1: The Framework

Process Component	Level A	Level B	Level C
II. Collect Data/ Consider Data	Understand that data are information; recognize that to answer a statistical investigative question, a person may collect data themselves specifically for that purpose, or a person may use data that have been collected by other people for another purpose Understand how to collect and record information from the group of interest using surveys and measurements collected from observations and simple experiments Understand that a variable measures the same characteristic on several individuals or objects and results in data values that may fluctuate Understand that within a data set there can be different types of variables (e.g., categorical or quantitative) Interrogate the data set to understand the context of the variables as they may relate to statistical investigative questions Understand that data are not always pristine but may contain errors, have missing values, etc., and that decisions have to be made about how to account for these issues	Understand that data are information collected and recorded with a purpose and can be organized and stored in a variety of structures (e.g., spreadsheets) Understand that a sample can be used to answer statistical investigative questions about a population. Recognize the limitations and scope of the data collected by describing the group or population from which the data are collected Understand that data can be used to make comparisons between different groups at one point in time and the same group over time Recognize that data can be collected using surveys and measurements, and develop a critical attitude in analyzing data collection methods Understand that quantitative variables may be either discrete or continuous Understand how to interrogate the data to determine how the data were collected, from whom they were collected, what types of variables are in the data, how the variables were measured (including units used), and the possible outcomes for the variables Understand that data can be collected (primary data) or existing data can be obtained from other sources (secondary data) Understand how random assignment in comparative experiments is used to control for characteristics that might affect responses	Word as: Apply an appropriate data collection plan when collecting primary data or selecting secondary data for the statistical investigative question of interest. Distinguish between surveys, observational studies, and experiments Understand what constitutes good practice in designing a sample survey, an experiment, and an observational study Understand the role of random selection in sample surveys and the effect of sample size on the variability of estimates Understand the role of random assignment in experiments and its implications for cause-and-effect interpretations Understand the issues of bias and confounding variables in observational studies and their implications for interpretation Understand practices for handling data that enhance reproducibility and ensure ethical use, including descriptions of alterations, and an understanding of when data may contain sensitive information Understand how concerns about privacy and human subjects may affect the collection and distribution of data Understand that in some circumstances, the data collected or considered may not generalize to the desired population, or this data may be the entire population

Table 1: The Framework

Process Component	Level A	Level B	Level C
III. Analyze the Data	Understand that the distribution of a categorical variable or quantitative variable describes the number of times a particular outcome occurs	Represent the variability of quantitative variables using appropriate displays (e.g., dotplots, boxplots)	Use technology to subset and filter data sets and transform variables, including smoothing for time series data
	Represent the variability of categorical variables or quantitative variables using appropriate displays (e.g., tables, picture graphs, dotplots, bar graphs)	Learn to use the key features of distributions for quantitative variables, such as: ◦ center: mean as a balance point, and median as the middle-ordered value	Identify appropriate ways to summarize quantitative or categorical data using tables, graphical displays, and numerical summary statistics, which includes using standard deviation as a measure of variability and a modified boxplot for identifying outliers
	Describe key features of distributions for quantitative variables, such as: ◦ center: mean as the equal share, and median as the middle-ordered value of the data ◦ variability: range as the difference between the greatest and least value, and dispersion as how many units from the equal share value ◦ shape: number of clusters, symmetric or not, and gaps	◦ variability: interquartile range and mean absolute deviation (MAD) ◦ shape: symmetric or asymmetric and number of modes Use reasoning about distributions to compare two groups based on quantitative variables Explore patterns of association between two quantitative variables or two categorical variables: ◦ measures of correlation: quadrant count ratio (QCR) ◦ comparison of conditional proportions across categorical variables	Summarize and describe relationships among multiple variables Understand how sampling distributions (developed through simulation) are used to describe the sample-to-sample variability of sample statistics Develop simulations to determine approximate sampling distributions and compute p-values from those distributions
	Recognize distributions can be used to compare two groups Observe whether there appears to be an association between two variables		Describe associations between two categorical variables using measures such as difference in proportions and relative risk
			Describe the relationship between two quantitative variables by interpreting Pearson's correlation coefficient and a least-squares regression line
			Use simulations to investigate associations between two categorical variables and to compare groups
			Construct prediction intervals and confidence intervals to determine plausible values of a predicted observation or a population characteristic

Table 1: The Framework

Process Component	Level A	Level B	Level C
IV. Interpret Results	Use statistical evidence from analyses to answer the statistical investigative questions and communicate results through structured answers with teacher guidance Make statements about the group or population from which the data were collected, recognizing that conclusions are limited to these groups and cannot be generalized to other groups Describe the difference between two groups with different conditions	Use statistical evidence from analyses to answer the statistical investigative questions and communicate results with comprehensive answers and some teacher guidance Acknowledge that looking beyond the data is feasible Generalize beyond the sample providing statistical evidence for the generalization and including a statement of uncertainty and plausibility when needed Recognize the uncertainty caused by sample to sample variability State the limitations of sample information (e.g., a sample may or may not be representative of the larger population, measurement variability) Compare results for different conditions in an experiment	Use statistical evidence from analyses to answer the statistical investigative questions and communicate results through more formal reports and presentations Evaluate and interpret the impact of outliers on the results Understand what it means for an outcome or an estimate of a population characteristic to be plausible or not plausible compared to chance variation Interpret the margin of error associated with an estimate of a population characteristic Acknowledge the presence of missing values and understand how missing values may add bias to an analysis Use multivariate thinking to understand how variables impact one another Communicate statistical reasoning and results to others in a variety of formats (verbal, written, visual) Understand how to interpret simulated p-values appropriately

Level A

Introduction

Essentials for each component

Example 1: Choosing the Band for the End of the Year Party – Conducting a Survey and Summarizing Data

Example 2: Family Size - Mean as Equal/Fair Share and Variability as Number of Steps

Example 3: What do Ladybugs Look Like – Collecting, Summarizing, and Comparing Data

Example 4: Growing Beans – A Simple Comparative Experiment

Example 5: Growing Beans (continued) – Time Series

Example 6: CensusAtSchool – Using Secondary Data and Looking at Association

Summary of Level A

Introduction

Students are surrounded by data. They may think of data as a tally of students' preferences such as favorite type of music, or as measurements such as length of students' arm spans or number of books in school bags. Level A students might be in elementary school, middle school or even higher grades, or they might be adults who are not enrolled in school; regardless of age, individuals should begin their study of statistics here.

Within Level A, students develop data sense—that is, an understanding that data are information. Students should learn that data are generated about specific contexts or situations. They learn that they can use data to answer statistical investigative questions about that context or situation. They also begin to learn how to interrogate data.

Students should have opportunities to generate statistical investigative questions about a specific context (such as their classroom) and determine what data might be collected or retrieved to answer these questions.

Students also should learn how to use graphical representations for their data, describe features of the distributions, and begin to use these descriptions in answering the posed statistical investigative questions.

Finally, students at Level A should develop informal ideas of how probability is connected to statistical reasoning. These ideas will help support them when they later use probability to draw inferences informally at Level B and more formally at Level C.

Level A students may collect data or may be presented with secondary data. Teachers should take advantage of naturally occurring situations in which students notice a pattern about some data and begin to raise questions. For example, when taking daily attendance one morning, students might note that many students are absent. The teacher could capitalize on this opportunity and have the students formulate statistical investigative questions that could be answered with attendance data.

Essentials for each component

Level A recommendations in the statistical problem-solving process are:

I. Formulate statistical investigative questions

→ Understand when a statistical investigation is appropriate

→ Pose statistical investigative questions of interest to students where the context is such that students can collect or have access to all required data

→ Pose summary (or descriptive) statistical investigative questions about one variable regarding small, well-defined groups (e.g., subset of a classroom, classroom, school, town) and extend these to include comparison and association statistical investigative questions between variables

→ Experience different types of questions in statistics: those used to frame an investigation, those used to collect data, and those used to guide analysis and interpretation

II. Collect/consider data

→ Understand that data are information; recognize that to to answer a statistical investigative question, a person may collect data themselves specifically for that purpose, or a person may use data that have been collected by other people for another purpose

→ Understand how to collect and record information from the group of interest using surveys, and measurements collected from observations and simple experiments

→ Understand that a variable measures the same characteristic on several individuals or objects and results in data values that may fluctuate

→ Understand that within a data set there can be different types of variables (e.g., categorical or quantitative)

→ Interrogate the data set to understand the context of the variables and how they may relate to statistical investigative questions

→ Understand that data are not always pristine but may contain errors, have missing values, etc., and that decisions have to be made about how to account for these issues

III. Analyze the data

→ Understand that the distribution of a categorical variable or a discrete quantitative variable describes the number of times a particular outcome occurs

→ Represent the variability of categorical variables or quantitative variables using appropriate displays (e.g., tables, picture graphs, dotplots, bar graphs)

→ Describe key features of distributions for quantitative variables such as:

   o center: mean as the equal share, and median as the middle-ordered value of the data

   o variability: range as the difference between the greatest and least value, and dispersion as how many units from the equal share value

   o shape: number of clusters, symmetric or not, and gaps

→ Recognize that distributions can be used to compare two groups

→ Observe whether there appears to be an association between two variables

IV. Interpret the results

→ Use statistical evidence from analyses to answer the statistical investigative questions and communicate results through structured answers with teacher guidance

→ Make statements about the group or population from which the data were collected, recognizing that these conclusions are limited to these groups and cannot be generalized to other groups

→ Describe the difference between two groups with different conditions

Example 1:
Choosing the Band for the End of the Year Party – Conducting a Survey and Summarizing Data

Formulate statistical investigative questions

Students at Level A may be interested in the favorite type of music among their peers. Imagine an end of the year party is being planned for a certain grade level, and there is only enough money to hire one musical group. A class from that grade level might pose the statistical investigative question:

What type of music do the students in our grade like?

This statistical investigative question attempts to measure a characteristic, type of music preference, in the population of students at the grade level. (Note that for the youngest or most novice students, it would make sense to investigate the question *What type of music do the students in our class like?* as the question posed above will require some inference from the class to the grade.)

Collect data/consider data

To answer the statistical investigative question, students need to collect data about the music they like. Before beginning data collection, however, it is important to think through data collection methods.

A survey is a natural data collection method for Level A. One possible survey question could ask: *What is your favorite type of music?* However, the survey question in this form could elicit many different responses, which might make it difficult to analyze the data. Following discussion of the pros and cons of an open-ended or more restricted question, students might amend the survey question to: *What is your favorite type of music: country, rap, or rock?* Because this question specifically asks respondents to choose among three options, it will be easier to manage and analyze the data. The downside to this question is that it restricts respondents' choices, so for someone who prefers jazz, their response will not indicate their favorite music.

Type of music is a categorical variable defined here by country, rap, or rock. The data that result from each child identifying their type of music preference are called categorical data.

Once students decide on a survey question, they could conduct a census in which every student in the class answers the survey question. At Level A, students should recognize that there will be individual-to-individual variability. As William Osler, a famous physician from the 1900s said, "Variability is the law of life, and as no two faces are the same, so no two bodies are alike, and no two individuals react alike and behave alike under the abnormal conditions which we know as disease." (Silverman, Murray & Bryan, 2008).

The analysis of the results from this one class will be used to infer what the favorite music type might be for the whole grade.

Suppose the class survey of 24 students in one of the classrooms generated the data shown in Table 1.

Table 1: Raw data collected

Name	Music
Aaron	Country
Aden	Rap
Alex	Rap
Angelica	Rock
Ana	Country
Ariella	Country
Eliana	Rap
Elizabeth	Rock

Name	Music
Emilio	Rap
Evangeline	Rock
Felicity	Country
Gabriel	Rap
Isabel	Rap
Jake	Rap
Jerry	Country
Leo	Country

Name	Music
Maria	Rock
Michael	Rap
Nat	Country
Penny	Rap
Sofie	Rap
Veronica	Country
Vicki	Rap
Xavier	Rap

Analyze the data

There are multiple ways to organize and represent the raw data. For instance, young children might create a bar graph by lining up according to their favorite type of music. Or they could use sticky notes on the board or floor to represent their category. Then they can count the number of students in each line or sticky notes in each category.

Level A students might also use a picture graph to represent the distribution of the categorical variable type of music. The distribution summarizes the data for the variable type of music by identifying the frequencies for each of the three categories. A picture graph uses a picture of some sort (such as a musical instrument) to represent an individual's preference. See Figure 1 for an example. Thus, each child who favors a particular music type would put a cut-out of the instrument directly onto the graph the teacher has created.

Instead of a picture of an instrument, another graphic representation—such as an X or a colored square—could be used to represent each individual preference.

Note the difference between a picture graph and a pictograph. In a picture graph an object such as a construction paper cut-out is used to represent one individual on the graph. A pictograph, on the other hand, uses a picture or symbol is used to represent several items that belong in the same category.

For example, on a pictograph showing the distribution of car riders, walkers, and bus riders in a class, a cut-out of a school bus might be used

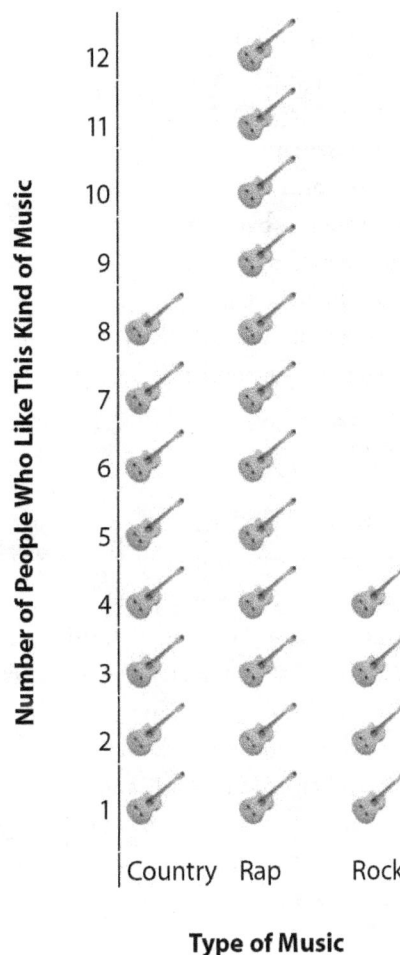

Figure 1: Picture graph of music preferences

to represent five bus riders. For example, if the class had 13 bus riders, there would be approximately 2.5 buses on the graph.

In either type of graph, if multiple symbols are used, it is important that they be the same size and be spaced the same distance apart to avoid visually distorting the data. For instance, if Figure 1 used pictures of a guitar, a microphone, and a drum, the pictures should be the same height and width.

Another common graphical display for categorical variables is a circle graph (pie graph). These graphs can be helpful for benchmark fractions of halves and quarters. Yet they also require an understanding of proportional reasoning and thus should only be used with students who have developed these skills.

The raw data from the music preferences survey can be summarized in the frequency table provided in Table 2. This frequency table is a tabular representation that summarizes the raw categorical data. Students might first use tally marks to track the categorical data before finding frequencies (counts) for each category.

Table 2: Frequency table of music preferences

Favorite	Frequency
Country	8
Rap	12
Rock	4

Level A students should also be introduced to bar graphs. A bar graph summarizes the data from some other representation, such as a picture graph or a frequency table. Figure 2 shows a bar graph of students' music preferences that were represented in the frequency table and picture graph. Note that because the data are categorical, the categories on a bar graph can be listed in any order.

Students at Level A should learn that the mode is the most common or most frequent outcome for a variable. The mode is a useful summary statistic for categorical data. Students should understand that the mode is the category that contains the most data points, which is often referred to as the modal category. If two categories are "tied" or contain the same number of data points, the distribution is bi-modal. Level A students should recognize the mode as a way to describe a "representative" or "typical" value for the distribution. In the favorite-music example, Rap music was preferred by more students than any other single music type; thus, the mode or modal category of the data set is Rap music. Students should see that the modal category (Rap) does not necessarily contain the majority of the student responses. In this example, there are 12 students who prefer Rap, but there are 12 students who do not prefer Rap.

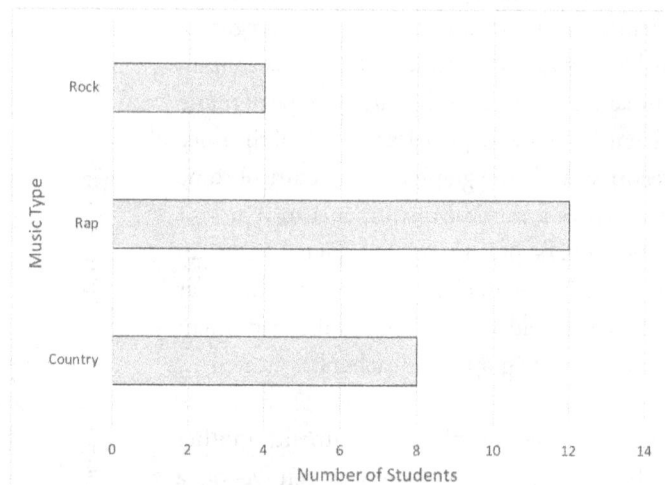

Figure 2: Bar graph of favorite type of music for students in class

Interpret the results

At this point in the investigation, teachers should urge students to answer the initial statistical investigative question:

What type of music do the students in our grade like?

A potential student answer might be:

The most popular type of music in our class was Rap. So we assume or infer that rap will be the most popular type of music in our grade.

Teachers should encourage Level A students to elaborate on such an answer to include consideration of the variability in the data. For example, the answer above could include the following description:

A total of 12 students preferred Rap, while only 8 preferred Country and 4 preferred Rock. There were 8 more students who preferred Rap rather than Rock.

If students have developed multiplicative reasoning, they might also say:

This shows that there were three times the number of students who preferred the most popular category compared to the number of students who preferred the least popular category.

At Level A, students should also begin to understand that probability is a measure of the chance that something will happen. It is a measure of the degree of certainty or uncertainty. The probability of events should be seen as lying on a continuum from impossible to certain, with less likely, equally likely, and more likely lying in between. Using these probabilistic notions, Level A students can elaborate on their answer to the statistical investigative question to include statements such as:

If a student in our class is selected at random, are they more likely be a student who prefers Rap, Country, or Rock?

Most of the data collected at Level A will involve a census of the students' classroom. So the first stage is for students to read and interpret what the data show about their class at a simple level. Recall that the original statistical investigative question asks about a certain grade level. Students at Level A should be encouraged to think about how well their class findings would be representative of other classes in their grade level and if the findings would "scale up" to this larger group.

Students at level B will collect data from larger groups, so in preparation, teachers at Level A should encourage students to think about groups larger than students in the classroom. These could include students in the whole school, in the school district, or in the state, or even all people in the nation. Students should note which variables (such as age or geographic location) might differ for the larger group.

In the previous music example, students might speculate that if they collected data on type of music preference from their teachers, the teachers might prefer a different type of music. They might also consider what would happen if they collected type of music preference from high school students in their school system.

Example 2: Family Size - Mean as Equal/Fair Share and Variability as Number of Steps

Formulate statistical investigative questions

Suppose a local school district is interested in knowing how many people live in the household of each student. Level A students might want to ask a similar question about how many people live in the household for students in their classroom. Teachers should guide the students to pose a precise statistical investigative question, one that incorporates the intended population (e.g., households of students in Mrs. Lopez's class), the variable to be measured (e.g., number of people in a household), and anticipates variability (e.g., asking about "typical household sizes," anticipates variability where as asking "the typical household size" suggests a deterministic answer akin to asking what is the average household size – note that this an analysis question, one we would ask when we do the analysis). One such example may be:

What are typical household sizes for students in Mrs. Lopez's class?

Collect/consider data

Suppose the teacher decides to work with nine students at a time in the classroom and asks each student, "How many people, including yourself, are in the household you lived in for most of the year?" This is the survey question to help answer the statistical investigative question posed earlier. Each student represents their family size with a collection of snap cubes.

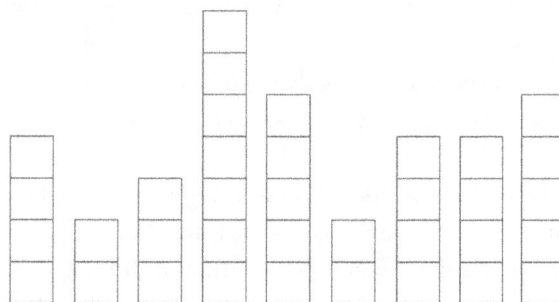

Figure 3: Snap cube stacks representing family size

The data for "family size" is represented with snap cubes as shown in Figure 3.

Analyze the Data

To examine the distribution of the household size for the collected data, students first arrange the stacks of snap cubes in increasing order (see Figure 4).

Students should recognize that family sizes vary. An analysis question that the teacher might ask the students is:

How many people would be in each family if all nine of these families were the same size?

When we make all the family sizes the same, the family size does not vary. Students may use two equivalent approaches:

(1) Disconnect all the snap cubes and redistribute them one at a time to the 9 students until all snap cubes have been

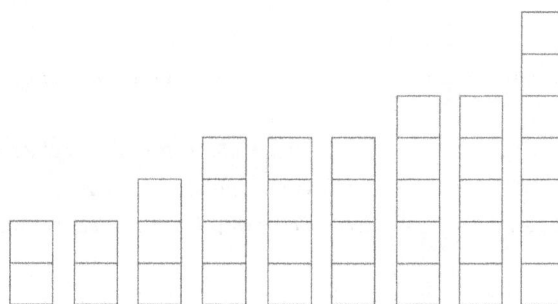

Figure 4: Ordered stacks representing family size

allocated. In this case, there are 36 snap cubes. Redistributing them among the 9 students yields 9 stacks with 4 snap cubes each.

(2) Remove one snap cube from the highest stack and place it on one of the lowest stacks, continuing until the stacks are leveled out.

Both methods yield an equal family size of four, which can be considered an equal share or a fair share.

For the second approach, students start by removing a snap cube from the highest stack and placing it on one of the lowest stacks. This yields a new arrangement of cubes. (See Figure 5.)

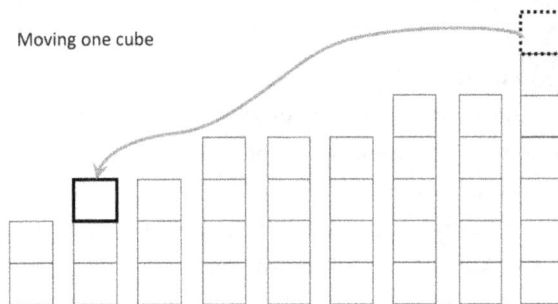

Figure 5: Moving one snap cube from the highest stack

Students continue this process until all the stacks are level, or nearly level when there is a remainder. (See Figure 6.)

After the final move, all nine stacks are level with four cubes each. This shows that the family size of four is an equal share. That is, if all nine family sizes were

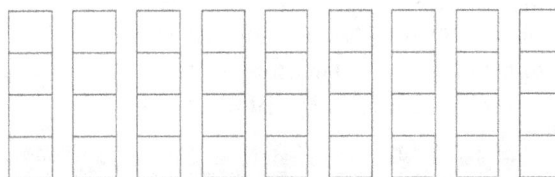

Figure 6: Snap cubes representing an average family size

the same, the number of people in the household would be four. This equal share value is the mean of the distribution. The mean is a numerical summary for the distribution of a quantitative variable used to measure the center of a distribution. The distribution of a quantitative variable summarizes the different values of the variable and the frequency with which they occur. Most students and teachers know how to calculate the mean by adding up all the observations and dividing by the number of observations. But what does the mean tell students about the distribution? How are students expected to interpret the mean? How are students expected to describe the variability in a distribution in relation to its mean?

The teacher might ask the students to investigate the following problem:

Suppose two other groups of nine students in the classroom found their equal share value to be six. What are some different snap cube representations that they could have constructed?

To answer this, students should first realize that they need to start with 54 snap cubes. They can then create two different distributions of family size where the equal share value is 6. For example, consider the following two groups, Group 1 (see Figure 7) and Group 2 (see Figure 8) of data on 9 family sizes from the classroom where the equal share family size for each group is 6.

Because the equal share value for each group is 6, the two groups cannot be distinguished based on the equal share value. An analysis question might be:

Which group is closer to being equal?

Students could offer different answers to this question, including:

(1) Group 2, because this group has the highest frequency of stacks of six snap cubes.

(2) Group 1, because for this group we would need to move fewer snap cubes to level out all the stacks to the equal share value of six.

The second method of having fewer snap cubes to move can be thought of as counting the 'number of steps to equal'—in other words, how many total steps would we need to move the snap cubes to create equal-sized groups. Fewer steps indicate that the distribution is closer to being equal and has less variability from the mean. Students can go through the process to see that for Group 1, they need to move eight cubes a total number of 48 steps. For Group 2, they need to move nine cubes a total number of 58 steps. Therefore, Group 1 is closer to equal and has less variability from the mean than Group 2.

Figure 7: Group 1 arrangement with average of 6

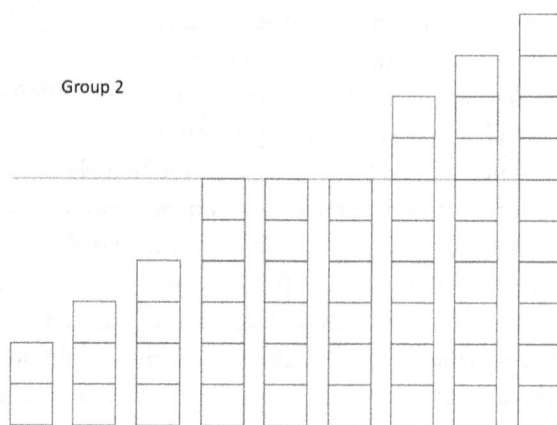

Interpret the results

Students should then interpret the results to answer the original statistical investigative question:

What are typical household sizes for students in Mrs. Lopez's class?

Using the results from the last two groups, students can comment that if all the families were the same size, the number of people in a household would be six, which would be the

Figure 8: Group 2 arrangement with average of 6

equal share or mean value. Group 1 has family sizes closer to the mean of six as measured by number of steps to level out the distribution. Group 1 has less variability than group 2. This illustrates that in looking at a distribution of household sizes, it is important not to rely upon one summary number (e.g., the mean). We also need to know how much household sizes vary from the mean to describe the typicalness of the sizes.

Level A students investigate and understand how to interpret the mean as the equal/fair share value and how to quantify variability from the mean as the number of steps from the equal share value. At level B, students will evolve to interpreting the mean as the balance point and to quantifying the variability from the mean as the mean absolute deviation (MAD).

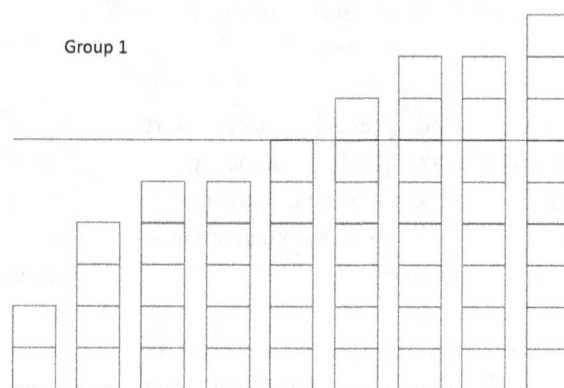

Example 3: What do Ladybugs Look Like – Collecting, Summarizing, and Comparing Data

Formulate statistical investigative questions

A common topic in science at Level A is to explore the structure and function of an organism's body parts and its environment. Current science standards include many statistical concepts and offer a place in the Pre-K–12 curriculum where statistical thinking can be introduced (https://ngss.nsta.org/Resource.aspx-?ResourceID=94). For example, consider a science unit in which students are studying what ladybugs do and what they look like (see Figure 9). With guidance and modeling from the teacher, the class formulates three statistical investigative questions to explore.

What does a ladybug usually look like?

How many spots do ladybugs typically have?

Do red ladybugs tend to have more spots than black ladybugs?

Collect data/consider data

Students are provided with secondary data in the form of pictures of ladybugs. As students view the picture cards, they may immediately notice that there is variation in the number of spots on the ladybugs and their color. Using the ladybug cards, the students can record information about the number of spots and the color of each ladybug pictured, or any other features they think might be relevant. Teachers guide students to create some data collection questions that will need to be answered for each ladybug in order to begin to answer the statistical investigative questions.

How many spots are on the ladybug?

What color is the ladybug?

What color are the spots on the ladybug?

The number of spots on a ladybug is an example of a numerical variable. We can obtain data on numerical variables from taking measurements (e.g., heights or temperatures of students) or counting objects (e.g., the number of letters in first names of students, the number of pockets on clothing

Figure 9: Hippodamia convergens (ladybug)

worn by students in the class, or the number of siblings each child has). Numerical variables are also called quantitative variables, which is the term used throughout this document.

The color of the ladybug is an example of a categorical variable. Data on categorical variables are observed according to their category, where the categories are mutually exclusive and jointly exhaustive, meaning they do not overlap and represent all possible observations.

With guidance from teachers, students select the categories to use for color: black, orange and red. For each of the photo cards (see Figure 10 for an example for 19 students), students will ask the data collection questions and record the information for the three variables: (1) color of the body, (2) number of spots, and (3) the color of the spots (if applicable).

Students should note that ladybugs are symmetrical, so if they count the spots on one side, then they know what the number is on the other side. The total number of spots is recorded. Sometimes data are messy or not clear. For instance, in this case, some of the spots are very faded and don't look like a spot at all. It is important that the class decide what will count as a spot (e.g., whether all shaped markings and all spots that are along the margin of the hard wing case will be counted). These consensus discussions will help reduce the measurement error introduced by the

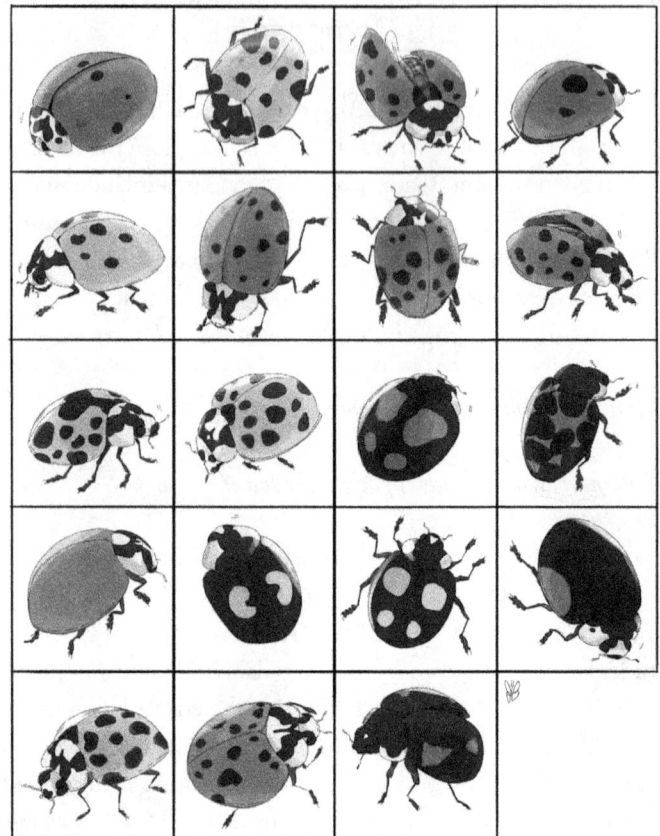

Figure 10: Examples of the 4x4 photo cards (see https://en.wikipedia.org/wiki/Harmonia_axyridis)

students recording information on the ladybugs they are viewing. Level A students should understand the importance of collecting data in a consistent manner.

Level A students might record the data they are collecting in a variety of ways. Early Level A students could consider one variable and record the values for each ladybug as in Table 3.

Table 3: Data table for early Level A

Ladybug #	Color of Body
1	R
2	O
…	…

Later in Level A, students might record all three variables at once. For example, they might record the answer to all three questions for each ladybug as in Table 4. This shows an example of a possible table structure showing from left to right in the cell: number of spots, color of body (R or B or O), color of spots (B or R or O). Each of the cells can be thought of as a data card, an organizational tool for data. Example 6 provides a detailed discussion of data cards.

Table 4: Example Table of ladybug data cards

6 R B	10 O B	16 R B	6 R B
10 O B	16 R B	18 R B	18 R B
20 O B	20 O B	4 B R	16 R B
0 R -	2 B R	4 B O	2 B R
20 O B	16 R B	4 B R	

By sharing their strategies as a class and discussing them, students can begin to recognize the importance of having a strategy that allows them to organize their data in a useful way.

Eventually, students should be looking to create a more productive way to organize the data. More experienced Level A students should be able to create a data table where each observation is on a separate row (see Table 5 as an example). This could be done on a worksheet using paper and pencil or using technology.

Table 5: Data table for ladybug cards

Ladybug #	Number of Spots	Color of Body	Color of Spots
1	6	R	B
2	10	O	B
...	...	...	...

Analyze the data

Early Level A students could use a picture graph to analyze the data. This allows them to keep track of which ladybug is being graphed. As the student advances, an appropriate graphical representation for one quantitative variable at Level A is a dotplot. With the teacher's guidance, students should be able to match a ladybug to a dot on their plot. This is an important connection, as a dotplot no longer allows an individual ladybug to be distinguished. Dotplots can be created by hand or using technology, and the horizontal axis typically represents the values of the variable under study. To compare the number of spots on ladybugs of different colors, students in Level A might use multiple dotplots with the same scale stacked one on top of the other. See Figure 11 as an example.

Using either a single dotplot or multiple dotplots broken down by the different colors of ladybugs, students can answer a series of analysis questions about the quantitative variable number of spots. For example, such questions might include:

What number of spots were most common/typical for all the ladybugs?
> *Red ladybugs only?*
> *Orange ladybugs?*
> *Black ladybugs?*

What is the least/greatest number of spots for the ladybugs?
> *Red ladybugs only?*
> *Orange ladybugs?*
> *Black ladybugs?*

Level A students should be supported to think about the distribution of a quantitative variable and the variability in the values. Students should understand that the median represents the value at the middle or center of the distribution of a quantitative variable. The same number of data points (approximately half) are greater than and are less than the median. The medians in Figure 11 are 14 spots for the red ladybugs, 18 for the orange, and 4 spots for the black.

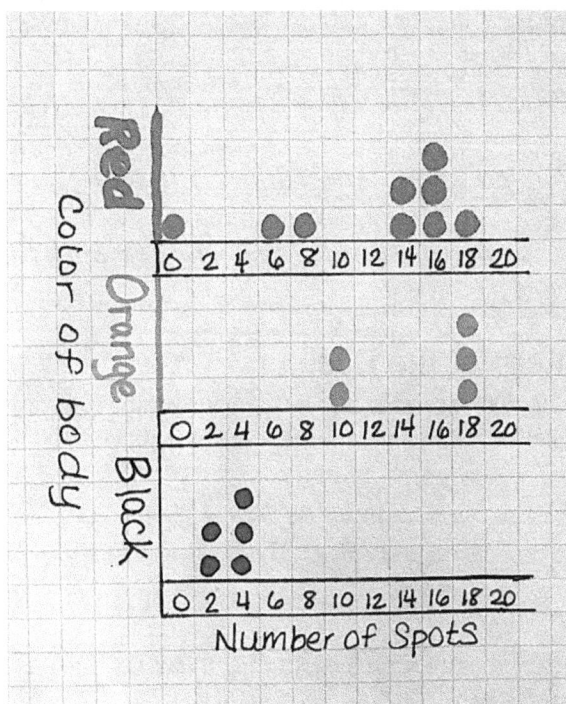

Figure 11: Drawn stacked dotplots of spot count for different color ladybugs

To illustrate the median as a center point, students can create a human graph to show how many spots are on the ladybug card they are holding. All the students with cards that have two spots stand in a line, with the students having cards with four spots standing in a parallel line next to them, then those with six spots, etc., until all students are lined up. Once all students are assembled, the teacher can ask one child from each end of the graph to sit down, repeating this procedure until one child is left standing, representing the median. Until students have mastered the idea of a midpoint, an odd number of data points should be used so the median is clear.

Analysis questions that require students to make assertions about the likelihood of certain statements allow Level A students to connect what they are seeing in their class to a larger scope. Students in Level

0		½		1
Impossible	Unlikely or less likely	Equally likely to occur as not occur	Likely or more likely	Certain

Figure 12: Scale of probabilities

A can learn to informally assign numbers to the likelihood that something will occur. An example of assigning numbers on a number line is shown in Figure 12.

Using these numbers, students in Level A could be asked the following analysis questions about the data at hand:

How likely is it to find a black ladybug that has more than five spots? Fewer than five spots?

For which color of ladybug did the number of spots vary the most? The least?

If I picked one of the ladybug cards out of a hat and it has six spots, what color do you think it will be?

Are orange ladybugs more likely to have 10 or more spots or less than 10 spots?

If I picked one of the ladybug cards out of a hat and it is a black ladybug, what chance is there that the number of spots it has is two?

Interpret the Results

Using their analyses, students can answer the statistical investigative question "What do ladybugs usually look like?" with an answer that might include the following:

Based on our pictures, the ladybugs in the card set were a mixture of red, orange and black ladybugs. The red ladybugs had between 0 and 18 black spots. The orange ladybugs had between 10 and 20 black spots. The black ladybugs were different, because they had either 2 or 4 spots in a mixture of colors.

A student may write the following to answer the question "How many spots do ladybugs in our card set have?":

Red ladybugs have between 0 and 18 spots. The most common number of spots is 16. The median number of spots for red ladybugs is 14 spots.

Orange ladybugs have between 10 and 18 spots. The most common number of spots is 18. The median number of spots for orange ladybugs is also 18 spots. This is a bit higher than the red ladybugs.

Black ladybugs have two or four spots. Out of the five ladybugs, two had 2 spots and the remainder had 4 spots.

To answer the comparative question, "Do red ladybugs tend to have more spots than black ladybugs?" students may answer:

Red ladybugs have between 0 and 18 spots. The most common number of spots is 16. The median number of spots for red ladybugs is 14 spots.

Black ladybugs only have 2 or 4 spots. The median of spots for black ladybugs is 4. There is only one red ladybug that has less than 4 spots; there is one that has zero spots. All other red ladybugs have 6 or greater number of spots. These analyses suggest that black ladybugs in our pictures tend to have fewer spots than red ladybugs.

Example 4: Growing Beans – A Simple Comparative Experiment

Another type of design for collecting data appropriate at Level A is a simple experiment, which consists of taking measurements on a condition or group. A simple comparative experiment is like a science experiment in which investigators compare the results of two or more conditions. For example, students might plant dried beans in two different growing conditions and let them sprout, and then compare which group grows fastest for the time period—the ones in the light or the ones in the dark. The data collected lead to a comparative experiment investigation. Students also might collect the growth data for their individual plants each day over the time period, which allows for a time series investigation.

Formulate statistical investigative questions
In the comparative experiment situation, a statistical investigative question might be:

After two weeks, do beans grown in the light tend to be taller than beans grown in the dark?

With comparison statistical investigative questions, the posed statistical investigative question is stated like a hypothesis (or conjecture) to indicate what the investigator believes will be the group that is bigger/larger/taller. In this case, the hypothesis is that beans grown in light will grow taller than beans grown in the dark.

Collect/consider data
Level A students will decide which beans will be grown in the light environment and which will be grown in the dark environment. Comparing these two conditions is the goal of the experiment.. The type of lighting environment is an example of a categorical variable. Measurements of the plants' heights, a continuous quantitative variable, can be taken at the end of a specified time period. These measurements can answer the statistical investigative question of whether one lighting environment is better for growing beans. When conducting experiments, the teacher needs to establish criteria with the students about how to handle certain situations that may arise. For example, some seeds may never grow, or certain plants may die, and students must decide how they will account for this in their data collection. The concept of an experiment is also more fully developed in Level C.

Analyze the data
Level A students can record the height of beans (in centimeters) that were grown in the dark and in the light using a dotplot. The heights on day 8 are represented in Figure 13.

Students make statements about what they notice in the dotplot. For example:

I notice that the median height for beans grown in the dark is 2.5 cm compared to the median height for beans grown in the light of 7.5 cm.

As students mature within Level A, the idea of the mean as an equal share can be introduced (see Example 2). The mean as a measure of center is developed further at Levels B and C. The mean and median are measures of location for describing the center of a quantitative variable.

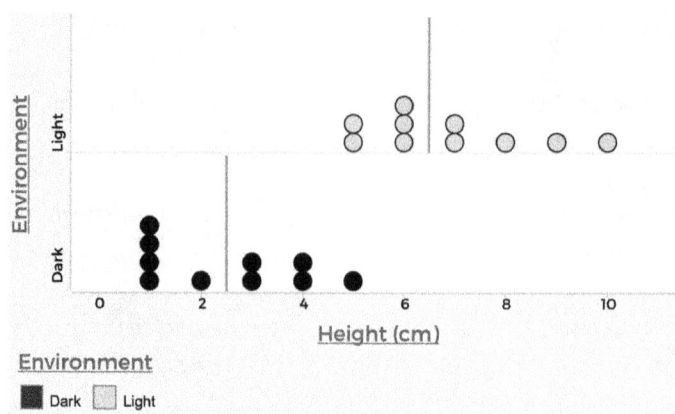

Figure 13: Stacked dotplots for bean height for light and dark conditions

It is also useful to know how the data vary across the horizontal axis. One measure of variability for a distribution is the range, which is the difference between the maximum and minimum values. Range only makes sense with data for a quantitative variable. The variability of the distribution also can be described by providing the lowest and highest values. For example, the heights of beans grown in the dark varies from 1 cm to 5 cm, whereas the beans grown in the light range in height from 5 cm to 10 cm. This is more informative for describing the distribution than just providing the ranges (for example, that the range of heights for beans grown in the dark is 4 cm and the range of heights for the beans grown in the light is 5 cm) because this provides measures of location in addition to variability.

Looking for clusters and gaps in the distribution helps students identify the shape of the distribution. Level A students should develop a sense of why a distribution takes on a particular shape for the context of the variable being considered.

→ *Does the distribution have one main cluster (or mound) with smaller groups of similar size on each side of the cluster?* If so, the distribution might be described as symmetric.

→ *Does the distribution have one main cluster with smaller groups on each side that are not the same size?* Students may classify this as "lopsided," or may use the term asymmetrical.

→ *Why does the distribution take this shape?* Using the dotplot from Figure 13, students will recognize both the beans grown in the dark and beans grown in the light have distributions that are "lopsided," with the main cluster on the left end of the distributions and a few values to the right.

As students advance to Level B, considering the shape of a distribution will lead to an understanding of which measures are appropriate for describing center and variability.

Interpret the results

The analyses reveal that the median height for the beans grown in the light exceeds the median height for the other beans by 5 cm. Nearly all of the beans grown in the light are taller than the beans grown in the dark. In fact, the beans that overlap are the shortest beans in the light condition and the tallest beans in the other condition. Thus, the plants from our experiment in the light environment tend to be taller than the plants in the dark environment.

Example 5: Growing Beans (continued) – Time Series

Formulate statistical investigative questions

Level A students can explore the height of their individual bean plants over a fifteen-day period. In this situation the statistical investigative question might be:

How does a bean plant that is in a light environment grow over the course of fifteen days?

Collect/consider data

For the comparative experiment, the students needed only to record the height at the end of fifteen days. To collect the data for time series, students record the height every day in order to track growth.

The data for one plant collected by one student can be recorded in a table (see Table 6).

Students can see from the table that day 9 is missing, and it is important to address this issue before analyzing the data. Why is there missing information and how might this affect the analysis of the data? The students were unable to access the plants on day 9 to take measurements. They decided that since they had six days of measurements afterward, the missing data would not affect their ability to see a possible trend.

Table 6: Bean plant grown in light, daily height

Bean Plant A, Growth in Light	
Day number	Height (cm)
1	0.75
2	1
3	1
4	2.2
5	4
6	6
7	6.5
8	7
9	N/A
10	10
11	10
12	12
13	13
14	14
15	16

Analyze the data

Level A students can use a time series graph called a timeplot to explore how a quantitative variable changes over time.

Using the timeplot pictured in Figure 14, students might notice:

The height of the bean plant has increased every day except from day 2 to 3 and from day 10 to 11.

The growth of the bean plant started quite slow. From day 1 to 3 it had hardly grown at all, less than a centimeter. The bean plant grew to 16 cm by the end of the two weeks.

The most the plant grew in one day was 2 cm. This amount of growth happened three times: between days 5 and 6, between days 11 and 12, and between days 14 and 15.

Students in Level A should be aware of how changing the scale of the vertical axis impacts perception of growth. They should see that increasing the spaces between the increments on the Height axis could make it appear that the beans grew faster than they did.

Interpret the results

The analyses show that the bean plant has grown from 0.75 cm to 16 cm in the fifteen-day period. The growth rate was fairly consistent across the fifteen days. Students can compare their results to others in the same growing condition. This way they can explore typical growth rates within each group and develop future statistical investigative questions.

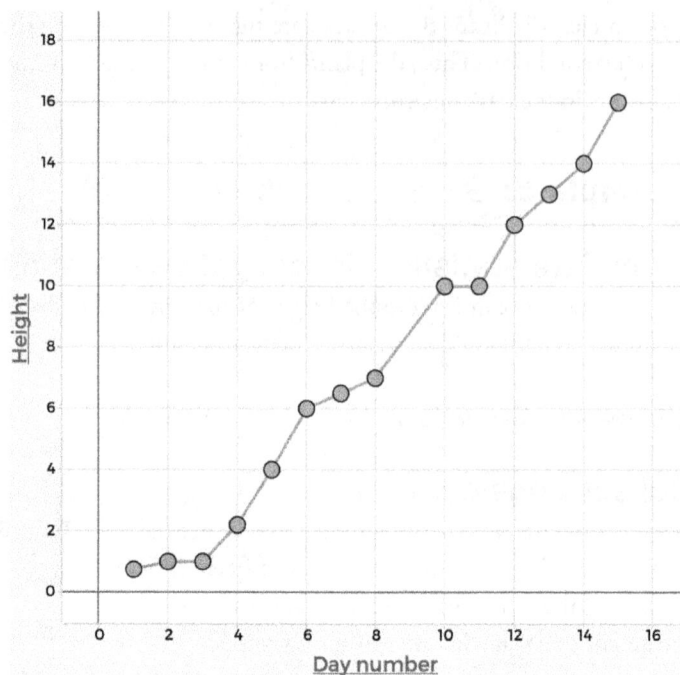

Figure 14: Timeplot of the height of a bean grown in light over two weeks

Example 6: CensusAtSchool - Using secondary data and looking at associations

When posing statistical investigative questions, instead of actually collecting data themselves, students may find a secondary data set to analyze and interpret in order to answer the questions. Secondary data are data that have been previously collected. The statistical investigative process can also begin by considering data directly. For example, a student may consider a secondary data set and then pose statistical investigative questions about the data set. This examples illustrates the latter scenario.

Collect data/consider data

Consider a secondary data set collected from Mr. Johnson's fifth-grade class at a local elementary school using the CensusAtSchool survey (www.amstat.org/censusatschool). CensusAtSchool is an international classroom project that engages fourth- through twelfth-grade students in statistical problem solving using their own real data.

The survey asked a series of 13 questions; five are chosen here for analysis. Each student's responses are presented on a separate data card. A data card is a card or paper that shows the values of the variables included in the data set for each individual in the data set. Data cards are one tool that allows students to visualize multiple recorded measurements on the same observational unit. Such tools can help students recognize that multiple attributes can be measured on one observational unit such as a student in a fifth-grade class. For example, Figure 15(a) shows the data card for one student and 15(b) shows the key for the data cards.

The card includes information on the main way the student gets to school (top of the data card), the height of the student in centimeters (left of the data card), the arm span length of the student in centimeters (right of the data card), the length of the right foot of the student in centimeters (bottom left value of the data card), and the time taken to get to school in minutes (bottom right value of the data card).

Students should examine the data card and realize the unit of measure for each variable. For example, students in the United States recognize the units are not standard but instead metric. These data cards include four quantitative variables and one categorical variable.

This particular data card is for a student who travels to school by "motor car" (i.e., car), is 138 cm tall, has an arm span (AS) of 133 cm and right foot length (RF) of 20 cm, and takes 5 mins to get to school (travel time). The data cards shown in Figure 16 are a subset of the whole data set and can be found in More 4 U along with instructions for preparation.

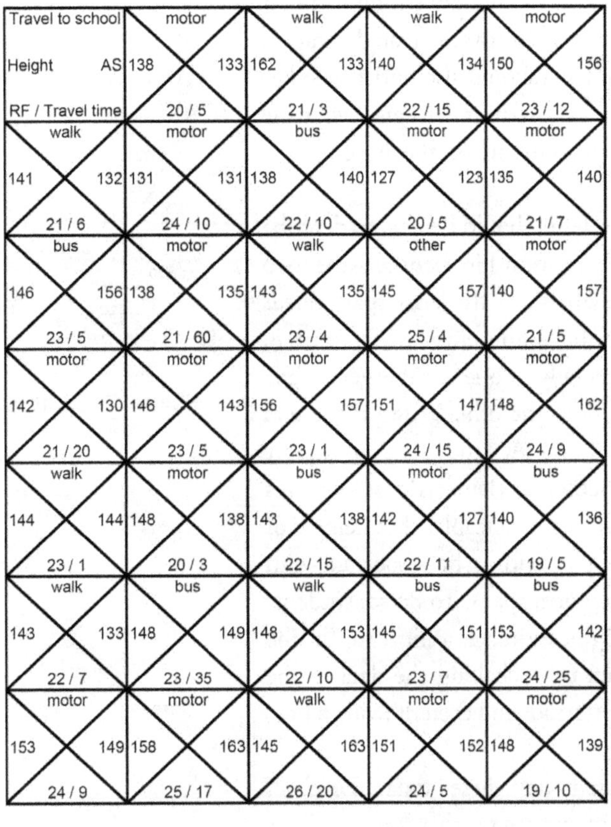

Once Level A students get the class set of data cards, they can interrogate the data set to identify and understand the context of each variable. Then they can pose interesting statistical investigative questions. More 4 U contains an example of how the data cards might be introduced in a way that encourages students to interrogate the data to identify the five variables for themselves.

Formulate statistical investigative questions

Several statistical investigative questions can be answered using these data, including the following:

Figure 15: (a) Example of a (a) student data card and (b) key showing positions of the five variables

Figure 16: 34 Data cards for Mr. Johnson's fifth-grade class

→ What are the heights of the students in this fifth-grade class? How long does it take students in this class who travel by bus to get to school? (summary statistical investigative questions)

→ Do the students in this fifth-grade class who travel by bus tend to take longer to get to school than the students in this class who walk to school? (comparison statistical investigative question)

→ Is there an association between height and arm span for the students in this fifth-grade class? (association statistical investigative question)

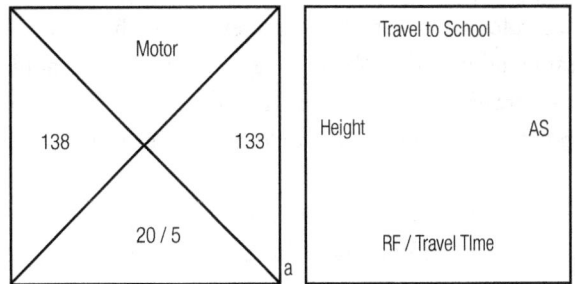

This statistical investigation will focus on the association statistical investigative question regarding arm span and height.

Analyze the data

A scatterplot can graphically represent data when values of two quantitative variables are obtained from the same individual or object. The focus at Level A should be on the interpretation of a scatterplot rather than its creation. Students can consult a constructed scatterplot of the heights and arm spans from the given data set (see Figure 17) to look for a relationship between these two quantitative variables.

Level A students can use the scatterplot to informally look for trends and patterns. They can question whether there is an association between height and arm span. If there is an association, they can note how strong it might be. For software-created graphs, students should also notice whether the axes start at zero and what the increments are on the graph.

Students should be able to describe the relationship between the two variables. In the arm span versus height scatterplot in Figure 18, generally, as one variable gets larger, so does the other. Toward the end of Level A, students can explore whether a claim is true for this class—for example, that heights of people tend to be associated with their arm spans. To do so, students might estimate a line to fit the data by eyeballing the data in the scatterplot and then drawing a line.

Based on these data, Level A students can see that some students have heights equal to arm spans; these are the students who are on the line, and 7 of the 34 students are on the line or close to the line. There are 10 students who have longer arm spans than heights; these students are above the line. Seventeen students are taller than their arm spans. Mostly the

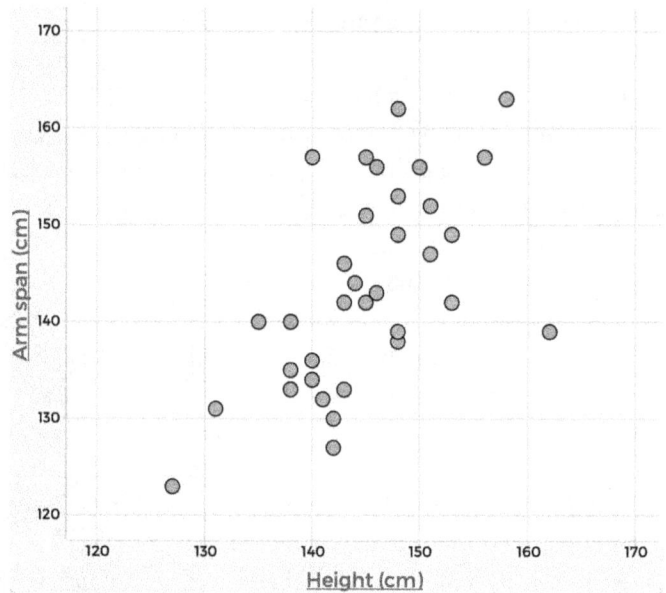

Figure 17: Scatterplot showing height and arm span

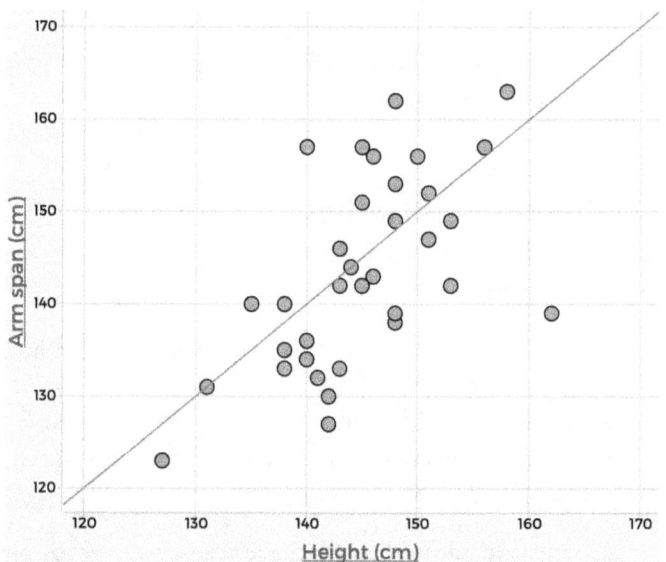

Figure 18: Scatterplot of height and arm span with a possible eyeball line

students' heights and arm spans follow the pattern of the line, suggesting that taller students have longer arm spans. One student has a very short arm span (approximately 139cm) compared to their height (approximately 162cm). Students in Level A may question whether the measurements were accurate for that student.

When students advance to Level B, they will quantify these trends and patterns with measures of association.

Interpret the results

The analysis reveals that most students in the class have heights that are similar to their arm span. Taller students tend to have longer arm spans. At Level A, students are exposed to situations that require the development of technical language to linguistically express relationships and patterns of data. By modeling appropriate use of summary terms such as "tend to, typical, usually, and similar," students can develop the technical language necessary to communicate their findings.

More 4 U provides an example where students at Level A can use data cards that include non-traditional items, which gives them early exposure to data science.

Summary of Level A

If students become comfortable with the ideas and concepts in Level A, they will be prepared to further develop and enhance their understanding of the key concepts for statistical literacy at Level B.

Through Level A statistical investigations, students begin to understand the role of questioning in the statistical problem-solving process, how to think about and represent multiple variables at a time, how to graphically display data in different ways, and how to think informally about probability. As students move from Level A to Level B to Level C, it is important to always keep at the forefront that understanding and managing variability is the essence of developing data sense.

Level B

Introduction

Essentials for each component

Example 1: Level A revisited: Choosing music for the school dance –
Multivariable and larger groups

Example 2: Choosing music for the school Dance (continued) –
Comparing groups

Example 3: Choosing music for the school dance (continued) -
Connecting two categorical variables

Example 4: Darwin's Finches - Comparing a quantitative
variable across groups

Example 5: Darwin's finches (continued) –
Separation versus overlap

Example 6: Darwin's Finches (continued) -
Measuring the strength of association between
two quantitative variables

Example 7: Darwin's finches (continued) –Time series

Example 8: Dollar Street – Pictures as data

Example 9: Memory and music - Comparative experiments

Summary of Level B

Introduction

Instruction at Level B should build on the statistical foundation developed at Level A and set the stage for statistical literacy at Level C. Instructional activities at Level B should continue to emphasize the statistical problem-solving process and have the spirit of genuine statistical practice. At Level B, students become more aware of how to use questioning in guiding their statistical reasoning. They also become attentive to the types of variables in data sets and how those variables can help address the statistical investigative questions posed.

Students continue to use their skills with graphical, tabular, and numerical summaries introduced at Level A. In Level B, they now broaden those skills to investigate more sophisticated problems, such as possible associations between variables. Level B students also shift from using data drawn from entire populations to working with data from samples of populations, drawing comparisons between two groups, and looking at changes over time. In Level B, the samples might not be random, so a focus at this stage is understanding the limitations and scope of their studies. Level B students also shift away from using smaller data sets and thus must begin to employ technology to help them navigate the statistical problem-solving process.

Essentials for each component

I. Formulate Statistical Investigative Questions

→ Recognize that statistical investigative questions can be used to articulate research topics and that multiple statistical investigative questions can be asked about any research topic

→ Understand that statistical investigative questions take into account context as well as variability present in data

→ Pose summary, comparative, and association statistical investigative questions about a broader population using samples taken from the population

→ Pose statistical investigative questions that require looking at a variable over time

→ Understand that there are different types of questions in statistics: those used to frame an investigation, those used to collect data, and those used to guide analysis and interpretation

→ Pose statistical investigative questions for data collected from online sources and websites, smartphones, fitness devices, sensors, and other modern devices

II. Collect/Consider Data

→ Understand that data are information collected and recorded with a purpose and can be organized and stored in a variety of structures (e.g., spreadsheets)

→ Understand that a sample can be used to answer statistical investigative questions about a population. Recognize the limitations and scope of the data collected by describing the group or population from which the data are collected

→ Understand that data can be used to make comparisons between different groups at one point in time and the same group over time

→ Recognize that data can be collected using surveys and measurements and develop a critical attitude in analyzing data collection methods

→ Understand that quantitative variables may be either discrete or continuous

→ Understand how to interrogate the data to determine how the data were collected, from whom they were collected, what types of variables are in the data, how the variables were measured (including units used), and the possible outcomes for the variables

→ Understand that data can be collected (primary data) or existing data can be obtained from other sources (secondary data)

→ Understand how random assignment in comparative experiments is used to control for characteristics that might affect responses

III. Analyze Data

→ Represent the variability of quantitative variables using appropriate displays (e.g., dotplots, boxplots)

→ Learn to use the key features of distributions for quantitative variables, such as:

    o center: mean as a balance point, and median as the middle-ordered value

    o variability: interquartile range and mean absolute deviation (MAD)

    o shape: symmetric or asymmetric and number of modes

→ Use reasoning about distributions to compare two groups based on quantitative variables

→ Explore patterns of association between two quantitative variables or two categorical variables

    o measures of correlation – quadrant count ratio (QCR)

    o comparison of conditional proportions across categorical variables

IV. Interpret Results

→ Use statistical evidence from analyses to answer the statistical investigative questions and communicate results with comprehensive answers and some teacher guidance

→ Acknowledge that looking beyond the data is feasible

→ Generalize beyond the sample providing statistical evidence for the generalization and including a statement of uncertainty and plausibility when needed

→ Recognize the uncertainty caused by sample to sample variability

→ State the limitations of sample information (e.g., a sample may or may not be representative of the larger population, measurement variability)

→ Compare results for different conditions in an experiment

Example 1: Level A revisited: Choosing music for the school dance – Multivariable and larger groups

Formulate statistical investigative questions

Students in Level A planned to select a musical group for an end-of-the-year party for a single grade by conducting a class census and answering the statistical investigative question:

What type of music do the children in our grade like?

Level A students initially collected data within their classroom to answer this question. Recall that the class was considered to be the entire population, and data were collected on every member of the population. Level A students then considered the plausibility of extending their class results to the entire grade.

A similar investigation at Level B might include planning for the entire school. This includes recognizing that one class may not be representative of the preferences of all students in one grade or all students at the school. Level B students might compare the preferences of their class with the preferences of other classes from their school and explore the following statistical investigative question:

What type of music do the students at our school like?

Collect /consider data

At Level B, students think about how they will collect and record data and who they can collect data from.

The statistical investigative question

What type of music do the students at our school like?

asks about the preferences of students at the school overall. In this case, a data collection plan could use a single class, such as a seventh-grade math class, as a sample to make decisions for the whole school. For this situation, students would discuss the limitations of their chosen sample. Alternatively, they might randomly select some students from each class or select two or three classes and get all the students in those classes to complete a survey.

Level B students should improve on survey questions used at Level A by understanding potential pitfalls to avoid in survey design (such as ambiguous wording and leading questions) as well as by providing more choices in answers. In addition, students should collect data on multiple aspects of a topic, which will foreshadow answering other statistical investigative questions.

For example, students can pose a series of survey questions that allow them to explore in more depth the types of music students like. After collecting the data, students can look at whether an association appears to be likely between different types of music they like. This information might inform the choice of music for the school dance.

Q1. Check yes for any of the following music types you like. Check no for any you don't like.

	Yes	No
Rap	☐	☐
Rock	☐	☐
Country	☐	☐
R&B	☐	☐
Pop	☐	☐
Classical	☐	☐
Alternative	☐	☐

Q2. What is your favorite type of music?

☐   Rap
☐   Rock
☐   Country
☐   R&B
☐   Pop
☐   Classical
☐   Alternative
☐   Other

Q3. What is your second favorite type of music?

☐   Rap
☐   Rock
☐   Country
☐   R&B
☐   Pop
☐   Classical
☐   Alternative
☐   Other

Q4. Would you prefer to have a live band or a DJ at the school dance?

☐   Live band
☐   DJ

As students move to Level B, they can design their data collection to use technology. Online survey tools offer ways to collect data and then download the resulting data into a spreadsheet. Having data accessible in a spreadsheet allows students to begin analyzing data using technology. They can also realize that when they complete surveys online, their data are collected in a central space for others to analyze and interpret.

Analyze the data

Many of the graphical, tabular, and numerical summaries introduced at Level A can be enhanced and used for more sophisticated analyses at Level B. Displays at Level B are likely to be representing multiple variables and/or using multiple displays to answer the statistical investigative question. To analyze the survey data collected using a class as a sample for the school, Level B students could graph the number of students who like each type of music (Figure 1). This bar graph uses student answers to survey question 1 noted above, where each music type is a variable.

The bar graph shows the frequencies of students who like and dislike each type of music. From this graph, students can see that rap has the highest number of students in the class responding

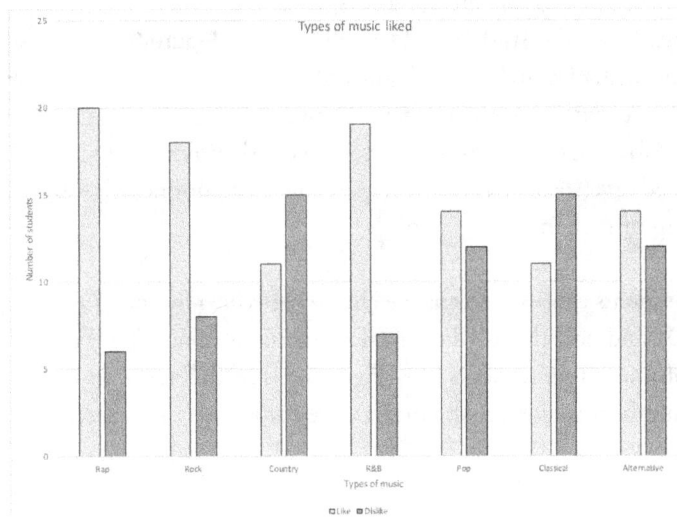

Figure 1: Side by side bar graph for liking music

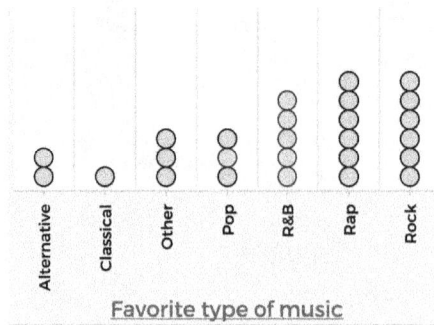

Figure 2: Favorite music by type

that yes, they like it (20 of the 26 students). R&B is second most liked, followed by rock. Pop and alternative also have more students who like it than dislike it. Country and classical have more students saying no than saying yes. The graph suggests that rap, R&B, and rock are the most popular music types for the class.

Responses to survey question 2 can be analyzed to see what students' favorite type of music is. Figure 2 shows there are an equal number of students (six) who identified rap and rock as their favorite music.

Students can explore favorite and second-favorite music (answers to questions 2 and 3 on the survey) through a two-way graph (see Figure 3). Here they can see that all but two students have rock, rap or R&B as their first and/or second choices for type of music preferences. In this graph, each dot represents a student in the class who responded to the survey. This two-way graph shows 49 bins (7 possible second favorite music types by 7 possible favorite music types). The bin in the top left corner has two dots in it, representing the two students in the class who answered that their favorite type of music is alternative, and their second favorite type of music is rock.

Analysis of the favorite and second-favorite types of music shows that nearly all the students (24 students, those shaded lighter in Figure 3) in the seventh-grade class have chosen R&B, rap, or rock as their first or second choice. Only two students (those shaded the darker color) did not rank R&B, rap or rock in the top two.

Students can also look at the choice between a live band and DJ and add this to their final conclusions about the types of music that students like (Figure 4). The difference in the number of students who prefer a live band versus a DJ is zero.

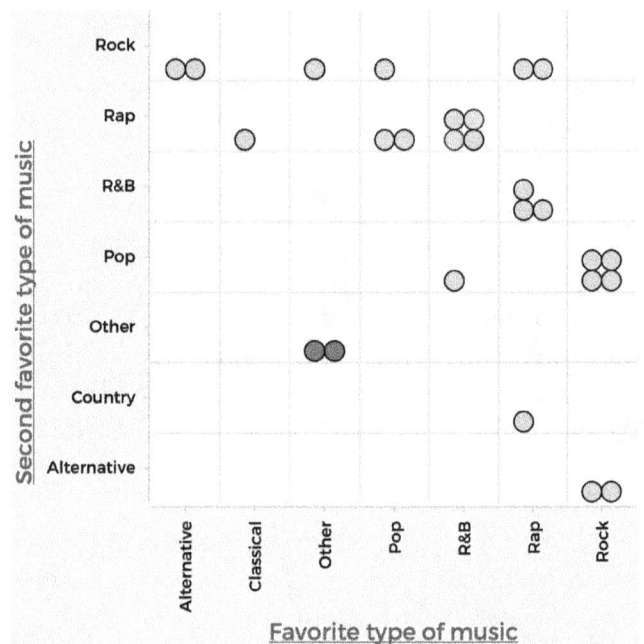

Figure 3: Favorite and second favorite type of music

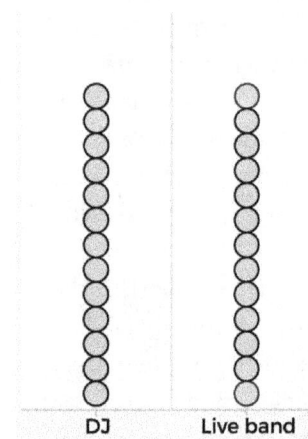

Figure 4: Choice of live band or DJ

Interpret results

The analyses show that R&B, rap, and rock are all very similar in terms of numbers of students who choose them as their favorite type of music. Altogether, 17 students in the class of 26 picked R&B, rap, or rock as their favorite type of music. In addition, Level B students might note that a live band and a DJ are equally chosen among class members.

Example 2: Choosing music for the school Dance (continued) – Comparing groups

Formulate statistical investigative questions

Level B students can consider the same data described above and compare them with a different group, such as a second class from another grade. They might consider the following statistical investigative question:

How do the types of music that students like differ between classes?

Collect /consider data

This statistical investigative question requires data to be collected from a second class (e.g., an eighth-grade math class) where the same survey questions can be used.

Analyze the data

As class sizes may be different, results should be summarized with relative frequencies or percentages to make comparisons. Level B students will use proportional reasoning to summarize and interpret data in terms of fractions and percentages.

The results from question 1 for the two classes are summarized in Table 1 using both frequency and relative frequency (percentages).

The tables show differences between the two classes. For example, whereas a majority of the seventh-grade class (Class 1) likes rock music (69%), a majority of the eighth-grade class (Class 2) does not like rock (68%). The bar graph in (Figure 5) compares the percentage of likes for each music category for the two classes.

Table 1: Frequencies and Relative Frequencies

Type of music	Class 1 Yes		Class 1 No		Total
Rap	20	77%	6	23%	26
Rock	18	69%	8	31%	26
Country	11	42%	15	58%	26
R&B	19	73%	7	27%	26
Pop	14	54%	12	46%	26
Classical	11	42%	15	58%	26
Alternative	14	54%	12	46%	26

Type of music	Class 2 Yes		Class 2 No		Total
Rap	21	75%	7	25%	28
Rock	9	32%	19	68%	28
Country	16	57%	12	43%	28
R&B	21	75%	7	25%	28
Pop	26	93%	2	7%	28
Classical	4	14%	24	86%	28
Alternative	10	36%	18	64%	28

Students at Level B should recognize that there is not only variability from individual to individual within a group but also variability from group to group. This second type of variability is illustrated by the fact that rap is the most popular type of music in Class 1, while pop is the most popular type of music in Class 2. That is, the modal category for Class 1 is rap music, while the modal category for Class 2 is pop music. From the comparative bar graph, Level B students should see that rap and R&B are similarly popular for Class 1 and Class 2; rock, classical, and alternative are less popular with Class 2; and country and pop are more popular for Class 2.

The results from the two classes might also be combined to create a larger sample (see Table 2). Combining the data across classes shows that the most popular type of music is rap with 41 votes followed by R&B and pop with 40 each.

Students might also consider results from the second survey question about favorite music type. Figure 6 combines the results for the two classes. Level B students should note that the favorite type of music is R&B (31%). Classical was the least popular choice for favorite type of music (2%). Rap (24%) and rock (19%) also stand out as preferred over the other types of music. Because there is a legend for the classes, it is easy to see that Class 2's overwhelming preference for R&B as their favorite type of music is what contributed to making this type of music the overall favorite for the two classes combined.

Interpret results

The analyses reveal that the most common types of music preferences for students at the school depends on the class. Class 1, which was a seventh-grade class, likes rap, R&B, and rock. Class 2, an eighth-grade class, likes pop, R&B, and rap. The favorite types of music for Class 1 are rap, rock, and R&B (all similar percents); for Class 2 the favorite type of music is R&B.

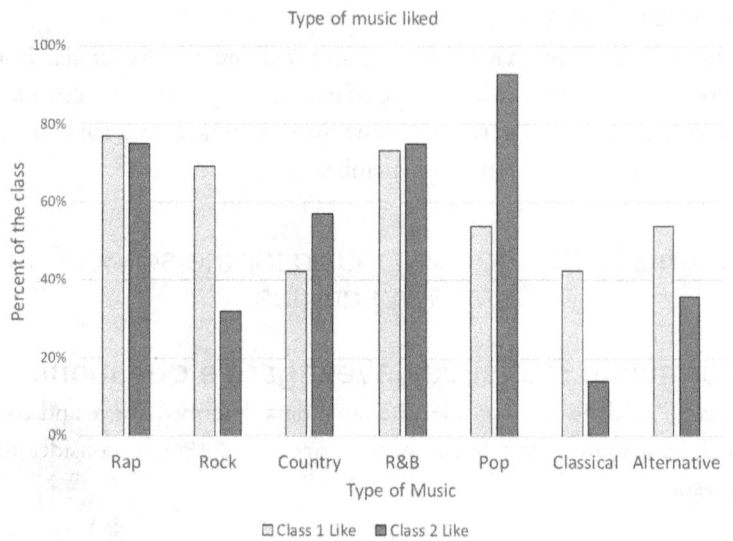

Figure 5: Comparative bar graph for type of music preferences

Table 2: Frequencies for Class 1 and Class 2 combined

Type of music	Class 1 & 2 Yes		Class 1 & 2 No		Total
Rap	41	76%	13	24%	54
Rock	27	50%	27	50%	54
Country	27	50%	27	50%	54
R&B	40	74%	14	26%	54
Pop	40	74%	14	26%	54
Classical	15	28%	39	72%	54
Alternative	24	44%	30	56%	54

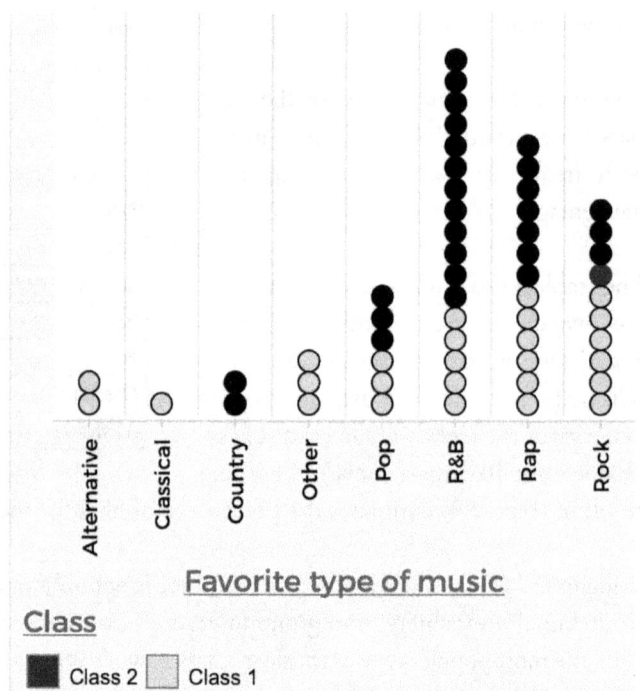

Figure 6: Combined favorite music type for the two classes

Because R&B had the greatest number of responses as the favorite music for both the eighth-grade class and the two classes combined, the students might argue for R&B music for the dance. However, more than 40% of those surveyed preferred either rap or rock music, so maybe a DJ who plays a mix of the three most preferred types of music would be best.

Level B students should recognize that although the combined sample is larger, it still may not be representative of the entire population (that is, all students at their school). In statistics, randomness and probability are incorporated into the sample selection procedure to provide a method that is "fair" and to improve the chances of selecting a representative sample. For example, if the class decides to select what is called a simple random sample of 54 students, then each possible sample of 54 students has the same probability of being selected. This application illustrates one of the roles of probability in statistics. Although students may not actually employ a random selection procedure when collecting data at Level B, issues related to obtaining representative samples should be discussed.

Example 3: Choosing music for the school dance (continued) - Connecting two categorical variables

Formulate statistical investigative questions

Level B students might pose additional statistical investigative questions that connect two categorical variables such as:

Do students who like rap music tend to like or dislike R&B music?

Do students who like rap music have a stronger tendency to like R&B music in comparison with students who do not like rap music?

These questions require Level B students to condition on one of the variables (Do students like Rap?) and consider ratios of students who like R&B across the two rap answer categories (yes/no). The first question considers only the "Yes" category to determine if these students like R&B more than they do not like R&B. In contrast, the second question begins to evaluate the association of two categorical variables and compares ratios for the two categories of "Yes" and "No."

Collect /consider data

Students can use the data collected from both classes to answer this statistical investigative question. Survey question 1 asks whether students like or dislike different types of music.

Analyze the data

Two categorical variables can be summarized in a two-way table (also called a contingency table). This table provides a way to investigate possible connections between two categorical variables. Such tables typically summarize the data using counts (frequencies). In this example, the categories are whether the students like (Yes) or dislike (No) rap music and whether the students like (Yes) or dislike (No) R&B music (see Table 3).

Table 3: Two-way table of likes/dislikes for rap and R&B

		Rap		
		No	Total	
R&B	Yes	36	4	40
	No	5	9	14
	Total	41	13	54

Alternatively, a two-way graph can be constructed to illustrate the data. In Figure 7 each dot represents a student in the class who responded to the survey.

This two-way graph shows four bins (two possible choices for rap, yes or no, by two possible choices for R&B, yes or no). The bin in the top left corner has 36 dots in it representing the 36 students who like rap and R&B. Level B students should be comfortable describing the manner in which the data are organized in two-way tables as well as noticing the benefits a visual representation can provide.

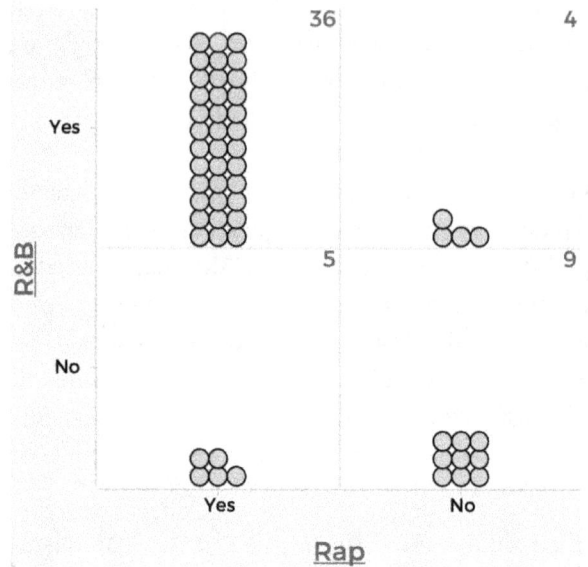

Figure 7: Two-way graph with counts

The contingency table and the two-way graph show that of the 41 students who like rap music, 36 also like R&B music. That is, 88% (36/41) of the students who like rap music also like R&B music. On the other hand, only 12% (5/41) of students who like rap do not like R&B. Alternatively, conditioning on the students who do not like rap, only 31% (4/13) of the students like R&B music. The percentage of all students who like R&B is 74% (40/54). Of the students who like R&B, a higher percentage of students like rap (88%) than those who do not like rap (31%). Level B students should notice the use of proportional reasoning in analyzing categorical data for two variables.

A mosaic plot can also be useful in considering the possible association of categorical variables (see Figure 8).

The mosaic plot visually shows that the proportions of people who like R&B across the two rap categories are clearly different. The proportions of the darker rectangles within the yes category and the no category of liking rap are vastly different. The yes/yes darker rectangle occupies 88% of the yes-rap column whereas the yes/no darker rectangle only occupies 31% of the no-rap column.

Level B students should understand that contingency tables can also be used to discuss probabilities, especially joint and conditional probabilities. The interior cells

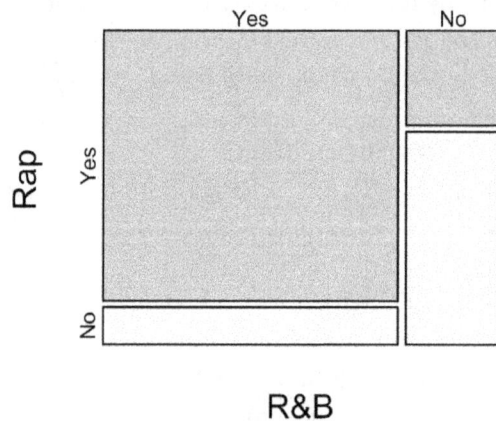

Figure 8: Mosaic plot for preferences of R&B and rap

provide the frequencies for joint events that are needed to calculate the probability of picking a student with a particular type of music preference at random from the group of 54 students. For example, the probability of selecting a student at random that likes both rap and R&B is 36/54. Conditional probabilities still use the joint frequencies, but they consider the marginal frequencies rather than the overall total. For example, the probability of selecting a student who likes R&B at random from those students who like rap is 36/41. Instead of conditioning on liking rap, conditioning in the other direction (on liking R&B) results in the probability of liking rap conditioned on liking R&B being 36/40.

Interpret results

The analyses indicate that students in this sample who like rap music also tend to like R&B. Of the students who like rap music, 88% like R&B music, whereas 12% do not like R&B music. Moreover, students who like rap music have a greater tendency to like R&B music in comparison to students who do not like rap music. The proportion of rap-liking students who like R&B is 88%, whereas the proportion of rap-disliking students who like R&B is only 31%, a difference of 57%. This difference suggests there might be an association between rap music and R&B music. It is important that Level B students understand there are limitations to generalizing these results beyond the classroom.

Example 4: Darwin's Finches - Comparing a quantitative variable across groups

At Level B, students will start to work with data that are not from their immediate environment. For example, as part of a science investigation, students can investigate Darwin's Finches.

Over the course of 2 million years, many species of finches in the Galapagos Islands have evolved from one common ancestor, the Grassquit. (Figure 9 shows 10 of the different species.) These finches have been the focus of research for scientists around the world since Darwin first observed them in the early 1800s. The different species have taken up different parts of the ecosystem in terms of where they spend most of their time and the types of foods they eat.

Within a given species, students in Level B can anticipate variation in the physical traits (phenotype), such as height, weight, beak length, and beak depth. This variation exists within the population because of the variation of gene expression within the species.

This example focuses on two specific ground finches: the Cactus Finch, *Geospiza scandens,* and the Medium Ground Finch, *Geospiza fortis.* The Cactus Finch primarily eats the fruit, seeds, nectar, and pollen from the prickly pear,

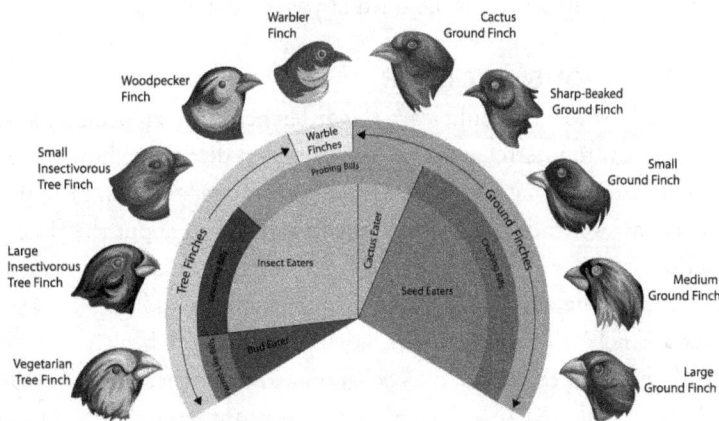

Figure 9: Finch species on the Galapagos Islands (https://stock.adobe .com/search?filters%5Bcontent_type%3Aphoto%5D=1&filters%5Bcontent _type%3Aillustration%5D=1&filters%5Bcontent_type%3Azip_vector%5D =1&filters%5Bcontent_type%3Avideo%5D=1&filters%5Bcontent_type%3 Atemplate%5D=1&filters%5Bcontent_type%3A3d%5D=1&filters%5B content_type%3Aimage%5D=1&k=darwins+finches+adaptive+radiation &order=relevance&safe_search=1&searchpage=1&searchtype =usertyped&limit=100&acp=&aco=darwins+finches+adaptive +radiation&get_facets=1)

Opuntia cactus. The Medium Ground Finch is considered a generalist and eats many different types of seeds. Their different eating habits might contribute to the evolutionary development of the shape of their beaks.

Formulate statistical investigative questions

Level B students can investigate the variation within a species and between species. In particular, they can explore the shapes of their beaks through beak length and beak depth (see Figure 10).

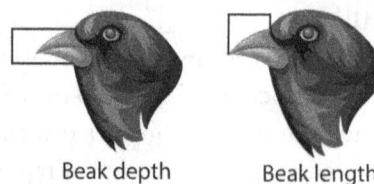

Beak depth   Beak length

Figure 10: Beak depth (mm) and beak length (mm)

Beaks are a fundamental feature of the bird that determines their ability to access food types and thus impacts their fitness. Students should be led to pose precise and interesting statistical investigative questions about finches' beak shape, such as:

Does either the Cactus Finch or Medium Ground Finch found on the Galapagos Islands tend to have a longer beak length than the other?

Level B students begin to understand what makes a good statistical investigative question. This question clearly states the variable of interest (beak length) and the populations of interest (Cactus Finches and Medium Ground Finches on the Galapagos Islands). In addition, the intention of the question is clear (comparing groups); the questions can be answered with the data available (there is information on the species and beak length in the data set); the question is interesting and worthwhile (it will help show possible differences between the two species of finches). These are all features that make this question a good statistical investigative question. A detailed discussion about questioning through the statistical problem-solving process is included at www. nctm.org/gaise.

Collect/Consider Data

At Level B, students should learn the difference between primary and secondary data. They should begin to learn how to interrogate secondary data and determine how the data were collected, how the variables were measured, what the units of observation are, and what the possible outcomes of the variables are. To help answer the statistical investigative question about the finches, secondary data can be used. The data set is a representative sample from a bigger data set[1] and contains data on 59 Medium Ground Finches and 42 Cactus Finches. The variables included in the data are beak length, beak depth, year of data collection, and whether the data were collected before or after a significant drought. Note that although the data set was created to have about 30 observations for each species from before the drought, and 30 observations for each species after the drought, there were fewer Cactus Finches data available after the drought.

Because the data are secondary, students should begin by exploring the data set. Looking at the summary table (Table 4), students can identify the variables included in the data set.

Table 4: Summary table of available variables

101 Total Finches	
Variable names	Description
Band	Id# of finch
Species	CF or MGF
Beak length	Length of beak in mm
Beak depth	Depth of beak in mm
Year	Year observation recorded
Drought	Before or after 1977 drought

---

1   See Grant, Peter R.; Grant, B. Rosemary (2013), Data from: 40 years of evolution. Darwin's finches on Daphne Major Island, Dryad, Dataset, https://doi.org/10.5061/dryad.g6g3h

The band represents the ID number of an individual finch. The species is a categorical variable with two categories Cactus or Medium Ground. The beak length is a continuous quantitative variable representing the measurement of the length from the base of the beak to the tip, measured in millimeters (mm). The beak depth is a continuous quantitative variable representing the height of the widest point in the beak, also measured in millimeters. The year variable is a discrete quantitative variable that identifies the year in which the measurements were taken. Lastly, the drought variable is a categorical variable with two categories indicating whether the measurements were taken before or after 1977, the year in which there was a major drought. Students in Level B should understand the difference between categorical and quantitative variables.

Students in Level B need to understand that in statistics, results are often generalized beyond the group studied to a larger group, the population. In this example, students are trying to gain information about the population by examining a portion of the population, called a sample. Such generalizations are valid only if the sample is representative of that larger group. As students evolve to Level C, the role of randomness in helping to provide representative samples will be developed and explored. A representative sample is one in which the relevant characteristics of the sample members are generally the same as those of the population. Improper or biased sample selection tends to systematically favor certain outcomes and can produce misleading results and erroneous conclusions.

Analyze Data

To answer the statistical investigative question, an initial comparative dotplot graph is drawn showing beak length split by species (see Figure 11).

This graph shows that the beak lengths for the sample of Medium Ground Finches (MGF) are approximately symmetrical and unimodal with the center just under 11 mm. The beak lengths for the sample of Cactus Finches (CF) are also approximately symmetrical and unimodal with a center at a little over 14 mm. The beak lengths for the MGF vary from 8.7 mm to 12.73 mm, and the beak lengths for CF vary from 12.8 mm to 16 mm (see Table 5).

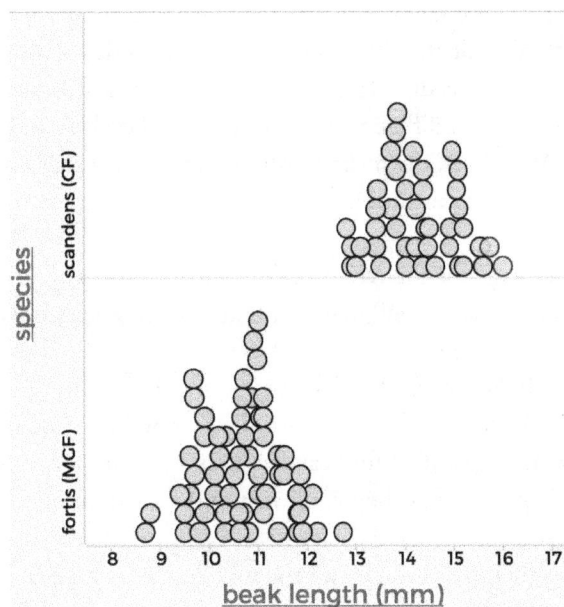

Figure 11: Comparative dotplot for Beak length

Problems that require comparing distributions for two or more groups are common in statistics. For example, at Level A, students might compare the number of spots on ladybugs of different colors or the height of beans grown under different conditions by examining parallel dotplots. At Level B, more sophisticated representations should be developed for comparing distributions. One of the most useful graphics for comparing quantitative data between two groups is the boxplot.

The boxplot (also called a box-and-whisker plot) is a graph based on a division of the ordered data into four quarters, with the same number of data values in each quarter (approximately one-fourth of the data). The four quarters are determined from the five-number summary (the minimum data value (Min), the first quartile (LQ), the median, the third quartile (UQ), and the maximum data value (Max)). The five-number summaries and comparative boxplots for the finch data are given in Table 5 and shown in Figure 12.

Table 5: Five-number summary for MGF (fortis) and CF (scandens) beak lengths all measured in millimeters

index	species	Min	LQ	Median	UQ	Max
1	scandens (CF)	12.8	13.7	14.2	14.94	16
2	fortis (MGF)	8.7	10.1	10.7	11.13	12.73

To interpret boxplots, students need to compare the global characteristics of each distribution (e.g., center and variability around the center).

For example, the median beak length for CF is 14.2 mm, which is 3.5 mm greater than the median beak length for MGF. The median suggests that the typical value for the CF beak length tends to be greater than the typical value for the MGF beak length.

Both the mean and median are measures of center. At Level A, the median was introduced as the quantity that has the same number of ordered data values on each side of it. This "sameness of each side" is the reason the median is a measure of center. The mean also takes on different interpretations in Levels A and

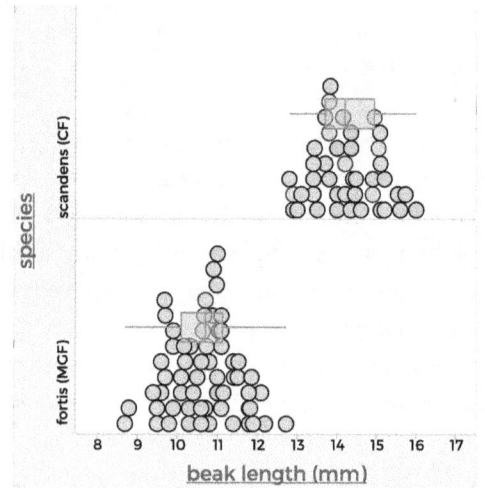

Figure 12: Comparative dotplots with boxplots for beak length

B. At Level A, the mean is introduced as equal share; at Level B the mean is used as a balance point (see www. nctm.org/gaise).

The range for the MGF beak length is 4.03 mm, versus 3.2 mm for the CF beak length. The ranges indicate that, overall, the distribution of beak lengths is more spread out for MGF than CF.

Another measure of variability that should be introduced at Level B is the interquartile range (IQR). The IQR is the difference between the first and third quartiles; it indicates the range of the middle 50% of the data. The IQR for beak length is 1.03 mm for MGF and 1.24 mm for CF. The IQRs show that the variability within the middle half of the distribution for MGF is slightly smaller than for CF.

Together, the median and the middle 50% of the boxplot can be used to make a visual determination as to whether the MGF beak length is greater/smaller than the CF beak length in their populations. Considering that the middle 50% of the CF data does not overlap with the middle 50% of the MGF data, it would be reasonable to say that in the general finch population, CF beak lengths tend to be greater than MGF beak lengths.

Interpret results

Because there is no overlap between the boxplots, these data suggest that on the Galapagos Islands, Cactus Finches tend to have longer beaks than Medium Ground Finches. Level B students should note that if they were to take another random sample, similar conclusions are likely.

Students in Level B also need to understand the notion of sampling variability—that is, how the values of a statistic (like the median, mean, range, IQR, etc.) vary from sample to sample. For example, one could take another sample of Medium Ground Finches on the Galapagos Islands at the same time and find the median length of their beaks might be different. At Level C, students will develop tools that provide a certain degree of confidence in the inference made from a sample to a population.

Example 5: Darwin's finches (continued) – Separation versus overlap

Formulate statistical investigative question

Students can further explore the finch data set by using the statistical problem-solving process to answer the statistical investigative question:

Does either the Cactus or Medium Ground finch species on the Galapagos Islands tend to have a greater beak depth than the other?

The teacher can encourage students to make a prediction about what they expect to see when they answer this statistical investigative question. With their prior knowledge about beak length, they can make an informed guess.

Collect/consider data

The same data set as described in the previous example can be used.

Analyze data

This statistical investigative question leads to a similar analysis of the data using comparative dot-plots and boxplots; however, the results are not as clear cut as they were in the previous analysis. Level B students should have the opportunity to wrestle with data that require more nuanced analysis. Students should reflect on the limitations of the data considered and the analyses conducted.

The initial comparative dotplot graph showing beak depth split by species is shown in Figure 13.

The graph shows that the beak depths for the sample of Medium Ground Finches (MGF) are approximately symmetrical and unimodal with a center just over 9 mm. The beak depths for the sample of Cactus Finches (CF) are approximately symmetrical and unimodal with a center higher than 9 mm. Level B students should understand that the peak of a distribution may not constitute the center of

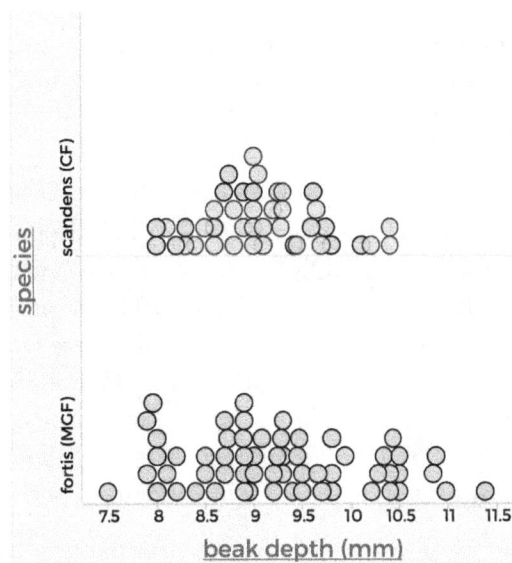

Figure 13: Comparative dotplots for beak depth

Table 6: Five-number summary beak depths

index	species	Min	LQ	Median	UQ	Max
1	scandens (CF)	8	8.63	9	9.44	10.4
2	fortis (MGF)	7.5	8.7	9.2	9.8	11.38

the distribution. The beak depths for the MGF vary from 7.5 mm to 11.38 mm, and the beak depths for CF vary from 8 mm to 10.4 mm (see Table 6).

An approach that works well with symmetric distributions is to describe the distribution using the mean as the measure of center and then develop a way to describe variability from the mean.

Since the distributions are not skewed or bimodal, the mean is an appropriate measure of center to use. The mean beak depth for the MGF is 9.26 mm, and the mean beak depth for the CF is 9.07 mm.

Variability can be described using the range and interquartile range. Variability can also be described by looking at how far the data values are from the mean. This measure of variability is called the Mean Absolute Deviation, or MAD. The MAD is the average distance of the data values from the mean. That is:

$$\text{"MAD} = \frac{\text{Total distance from the mean for all values"}}{\text{(Number of data values)}}$$

The MAD for the MGF is 0.73 mm, and the MAD for the CF is 0.49 mm. This indicates that individual beak depths for the MGF differ from the mean of 9.26 mm by 0.73 mm on average. The individual beak depths for the CF differ from the mean of 9.07 mm by 0.49 mm on average. The MAD also serves as a precursor to the standard deviation, which is developed at Level C.

A detailed investigation exploring the mean and the MAD is included at www. nctm.org/gaise.

In the same spirit as the first statistical investigative question, comparative boxplots are also drawn. Figure 14 shows there is a lot of overlap in the boxplots.

The range of the middle 50% of beak depths measured by the IQR for the MGF is 1.1mm, with depths varying from 8.7 mm to 9.8 mm, and the range of the middle 50% of beak depths for CF is 0.81 mm, with depths varying from 8.63 mm to 9.44 mm. The box parts of the boxplots of the sample data overlap almost completely, and the medians both lie within the overlap.

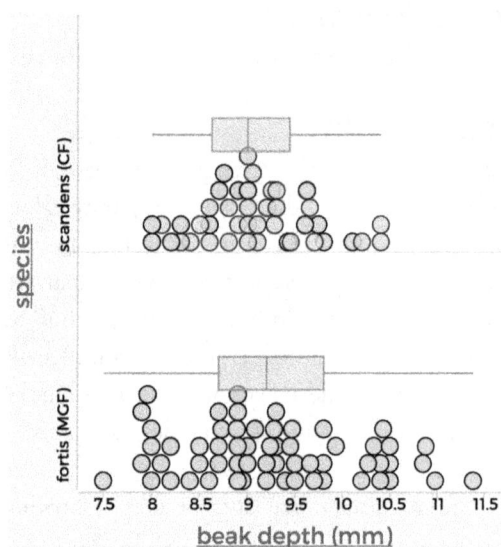

Figure 14: Comparative boxplots for beak depth for Cactus (scandens) finch and Medium Ground (fortis) finch

Interpret results

There is no obvious indication as to whether the beak depth of the Medium Ground finches on the Galapagos Islands is greater than or less than the beak depth of the Cactus finches on the Galapagos Islands. The mean beak depth differs by 0.19 mm, which is relatively small for the data given.

When results are not immediately obvious, students in Level B must acknowledge the ambiguity by noting sample-to-sample variation. If repeated samples of the finches were taken, the mean beak length can be found from each sample. Some variability within these samples is expected. A different sample might have a mean beak depth for the CF that is greater than the mean beak depth for the MGF.

Incorporating randomness in the sampling procedure allows the use of probability to describe the long-run behavior in the variability of the different sample means. Suppose each student took a random sample of finches and computed the sample mean. The students could then compile their sample means to create a class distribution of the sample means. Students can begin to make informal probability statements about which values of the sample mean are likely and which values are unlikely by observing how their sample means vary from the lowest to the highest value in the class. The variation in results from repeated random sampling is described through what is called the sampling distribution. The sampling distribution behavior allows the quantification of how close a sample mean is expected to be to the actual population mean. Sampling distributions are explored in more depth at Level C. The online resources found at www. nctm.org/gaise also include an introductory investigative task that explores random sampling, sampling distributions, and sampling variability through simulations that are appropriate for transitioning from Level B to Level C.

Example 6: Darwin's Finches (continued) - Measuring the strength of association between two quantitative variables

At Level B, more sophisticated data representations should be developed for the investigation of problems that involve the relationship between two quantitative (numerical) variables. The finch example can be extended for this purpose.

Formulate statistical investigative questions

An interesting observation that students may make from looking at the pictures of finches is that the depth and length of a beak might be associated in some way. A statistical investigative question to explore might be:

Is there an association between the depth and length of the beak for the Medium Ground finch?

Collect/consider data

The same data set described in the previous examples can also be used to answer this statistical investigative question.

Analyze results

To analyze the association of two quantitative variables, students at Level B should construct a scatterplot. This can be done by hand with small data sets or with technology for larger data sets. The beak length and beak depth data for Medium Ground finches are displayed in Figure 15. Similar to Level A, students at Level B should note when the horizontal and vertical axes do not start at zero and how the increments on the axes as these can impact the visual pattern. For example, compare Figure 15a, where the axes start at the

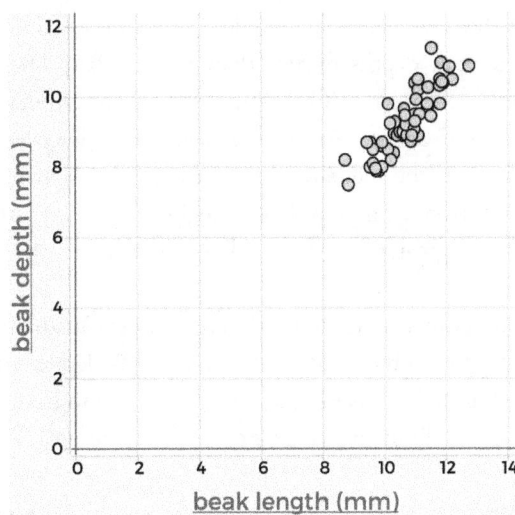

Figure 15a: Beak length and beak depth for Medium Ground Finch with zero origin

origin, with Figure 15b, where the axes start at the software-determined default. The scatterplot in Figure 15b suggests a positive relationship between beak length and beak depth; as the beak length increases, so does the beak depth. In addition, the relationship appears to be quite linear.

Measuring the strength of association between two variables is an important statistical concept that should be introduced at Level B. This concept is developed further in Level C and beyond; however, the Quadrant Count Ratio (QCR) can be introduced to measure association at Level B.

To help students visually identify patterns in the distribution of points within a scatterplot, the two lines divide the scatterplot into four regions (or quadrants). The scatterplot in Figure 16 for the beak length/beak depth data includes a vertical line drawn through the median beak length ($x = 10.7$ mm) and a horizontal line drawn through the median beak depth ($y = 9.2$ mm). It should be noted that these lines could also be placed at the mean beak length and mean beak depth (see at www.nctm.org/gaise) instead of the medians.

The upper-right region (Quadrant 1) contains points that correspond to finches with beak lengths greater than the median and beak depths greater than the median. The upper-left region (Quadrant 2) contains points that correspond to finches with beak lengths less than the median and beak depths greater than the median. The lower-left region (Quadrant 3) contains points that correspond to finches with beak lengths less than the median and beak depths less than the median. The lower-right region (Quadrant 4) contains points that correspond to finches with beak lengths greater than the median and beak depths less than the median.

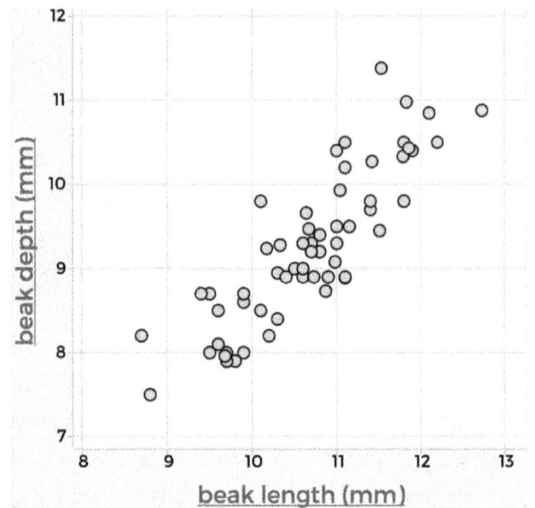

Figure 15b: Beak length and beak depth for Medium Ground Finch without zero origin

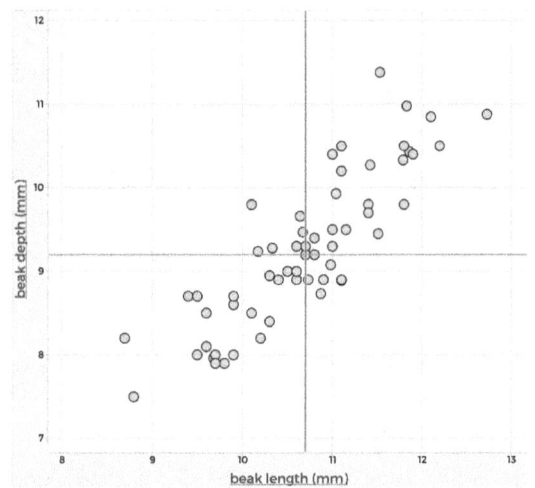

Figure 16: Scatterplot of beak length and beak depth with medians indicated by the lines

Most points in the scatterplot are in either Quadrant 1 or Quadrant 3. That is, most MGFs with beak lengths greater than their median also have beak depths greater than their median (Quadrant 1) and most MGFs with beak lengths less than their median also have beak depths less than their median (Quadrant 3). These results indicate that there is a positive association between the two variables beak length and beak depth.

Generally stated, two quantitative variables are positively associated when above average values of one variable tend to occur with above average values of the other, and when below average values of one variable tend to occur with below average values of the other. Negative association between two quantitative variables occurs when below average values of one variable tend to occur with above average values of the other.

A measure of correlation is a quantity that measures the direction and strength of an association between two quantitative variables. Level B students should note that the points in Quadrants 1 and 3—here, a total of 44 points—contribute to the positive association between beak length and beak depth. Points in Quadrants 2 and 4—here, a total of 10 points—contribute to the negative association between beak length and beak depth.

One measure of correlation between beak length and beak depth is given by the QCR (Quadrant Count Ratio):

$$QCR = \frac{(n_{Q1}+n_{Q3}) - (n_{Q2}+n_{Q4})}{n} = \frac{(44-10)}{54} = 0.63$$

The QCR is found by taking the difference between the number of points that are consistent with a positive association and the number of points that are consistent with a negative association, then dividing by the total number of points. The QCR is unitless and is always between −1 and +1, inclusive.

Level B students should be encouraged to look for overall trends in a scatterplot. If the trend is linear, students may draw a line on the scatterplot that they believe best describes the trend. At this level, this should be done in an exploratory fashion. Students study linear relationships in other areas of the mathematic curriculum, which provides a good opportunity to connect algebra and statistics. The degree to which these ideas have been developed will determine how students can proceed in their exploration.

Interpret results

The scatterplot reveals a positive linear association between the beak length and the beak depth. The QCR of 0.63 also shows that there is a fairly strong positive association between the two variables. These results suggest that the MGF beak length is positively associated with the MGF beak depth. At Level C, the QCR is used as a foundation for developing Pearson's correlation coefficient. Level C also introduces more formal ways of exploring linear relationships between two quantitative variables. For more discussion about the advantages and shortcomings of the QCR as a measure of correlation, see Kader & Franklin, 2008. At www.nctm.org/gaise, measures of association for categorical variables are presented.

Example 7: Darwin's finches (continued) –Time series

Another important type of statistical investigative question that should be considered at Level B is a question that requires students to explore trends in data over time. Data for investigating trends over time are known as time series and are quite common in the sciences.

Formulate statistical investigative questions

The finch example can be extended to a time-related statistical investigative question:

How does the average length of the beak of the Medium Ground finches shift over time?

Collect/consider data

Within the overall finch data set, there are subsets of data that can be used to answer this statistical investigative question. See Grant, Peter R.; Grant, B. Rosemary (2013), Data from: 40 years of evolution. Darwin's finches on Daphne Major Island, Dryad, Dataset, https://doi.org/10.5061/dryad.g6g3h (Fig. 01-06 (also 7.3)).

Figure 17 shows that the two variables needed to answer the statistical investigative question (year and average beak length) are available in the data set.

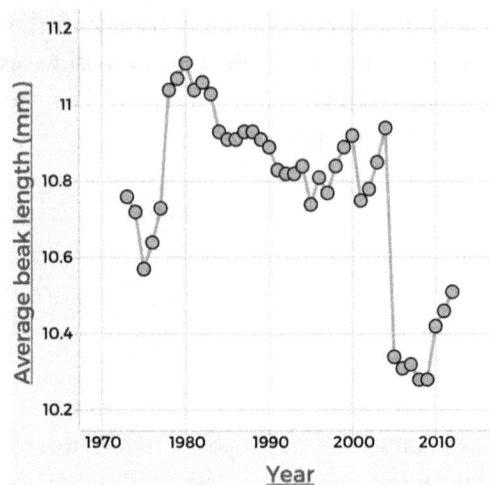

Year	1973–2012
Species	fortis,
Average beak length	10.28–11.11 mm
Average beak depth	8.51–9.81 mm
Average beak width	8.27–8.93 mm
CI Beak length	0.05–0.22
CI Beak depth	0.05–0.22
CI Beak width	0.04–0.17

Figure 17: Variables available in data set

Figure 18: Time series graph of the average beak length for Medium Ground finches from 1973 to 2012

Analyze data

To examine the change in average beak length over time, a time series plot can be used (see Figure 18). Students at Level B should reason that time can be plotted on the horizontal axis and the other variable of interest on the vertical axis. Much in the spirit of a scatterplot, students can plot the lengths recorded over time. To better show general trends and patterns over time, it is appropriate to connect the points for a time series graph.

Interpret Results

A student examining the time series plot might notice that the average beak lengths for MGF vary over the 35-year period. Initially, they start at 10.75 mm, drop until 1975, and then rise until a peak in 1980. After 1980, average beak length shows a general trend of slow decline, including a big drop between 2004 to 2005. Finally, beak length started to increase again in 2009. Students can research the Medium Ground finches on the Galapagos Islands to see if they can find explanations for the features of the time series, especially the sharp rises and the large drop. Students might explore questions such as: Was there a change in food available in those years? Were the number of finches the same in each data collection period? At Level B, students can describe what they see. At Level C, students move from describing data into making predictions.

Example 8: Dollar Street – Pictures as data

As students enter Level B, they should begin to understand that data are all around them. Data can be used to understand the world, including pressing issues that are relevant to the world's population. Students at Level B should begin to understand that the statistical problem-solving process allows them to engage with large and complex data sets to understand better the world around them. Level B students also need to understand that the concept of data also includes pictures, text, sound, etc. An important exercise for students at Level B is to become familiar with navigating different data types, not just those stored in static worksheets. Such exercises at Level B focus on students trying to make sense of non-traditional data.

Formulate statistical investigative question

The use of photographs as data coupled with interview data provides an opportunity to challenge misconceptions of families, cultures, and countries, and to realize that our daily lives are often more similar than different. Anna Rosling Ronnlund, the developer of the interactive Gapminder website called Dollar Street, states *"People in other cultures are often portrayed as scary or exotic. This has to change. We want to show how people really live. It seemed natural to use photos as data so people can see for themselves what life looks like on different income levels. Dollar Street lets you visit many, many homes all over the world. Without travelling."* At the time of this writing, the Dollar Street project has taken pictures of 300 homes across 56 different countries. Over 30,000 photos are included in the data set. The photos taken and the background of families living in these homes can be accessed at:

https://www.gapminder.org/dollar-street/about

Using the picture data available, students in Level B may explore answers to the following statistical investigative question:

How are people's concepts of family and living spaces similar or different across the world?

Collect data/consider data

Pictures as data are increasingly common as picture taking is now more accessible through phones, and as people are habitually taking pictures of their daily happenings. Level B students can consider non-conventional data such as picture data through questioning and exploration. To answer the posed statistical investigative question, students in Level B need first to interrogate the Dollar Street data, which is a secondary data set, to understand it better. Exploring these types of data sets, students should begin to wonder and research how and why the data were collected, from whom the data were collected, how variables were measured, and what were the possible outcomes for the variables.

How the data were collected. In accordance with sound ethical guidelines, all the families participating in Dollar Sense are volunteers. Various photographers collected data and spent a day in each family's living space across the 56 countries. At each living space, the photographers took pictures of up to 135 objects that were then tagged and included in the data set. In addition, the photographers interviewed families with the same set of questions across the world. The questionnaire provides context for the pictures. Families participating in Dollar Street are volunteers, and families may be added to the data set at any time. To add a family to the data set, a consent form must be signed by the family and a background questionnaire is completed. The questionnaire can be found here:

https://docs.google.com/document/d/1vtracv6xSEDWvglYVDi7k2fAATs55wZUwnMdECIIfS4/edit#heading=h.x1v8rup1z5m0

From whom were the data collected? Data were collected from families across the world. A family is considered a unit of observation (case) in the data set. For each family, numerous variables were recorded, making the data multivariable.

Variables collected. There are over 150 variables defined in the data set, including beds, toothbrushes, front doors, work areas, bathrooms, and inside walls. The variables are recorded by the pictures themselves. In addition to the variables captured by the pictures, the dataset also contains a variable defining an estimate of the monthly consumption per adult living in the living space, expressed in terms of dollars spent. For information on the assumptions, limitations, and and an explanation of how the monthly consumption value was calculated, see Lindgren, https://drive.google.com/drive/folders/0B9jWD65HiLUnRm5ZNWlMSU5GNEU.

Analyze data

Level B students should be given generous time to explore these data. Often with large data sets and non-traditional data sets, students need time to digest and merely "play" with the data. The data are so rich that it is difficult to parse through the number of variables and nuances of the images in a short time frame. In addition, much of the value from working with non-traditional data lies in honing questioning skills. At the outset of the investigation, students can be encouraged to record several observations and wonderings about the images or the site. These observations and wonderings will help students become familiar with the site layout, the complexity of the data set, and allow teachers to address generalizations or biases that students might be unintentionally using. It is useful at this point to remind students that the intent of the Dollar Street data set was to quantify consumption by exploring families' available spending money.

After exploration time, the teacher can guide students in Level B to select a variable such as the Family variable shown in Figure 19.

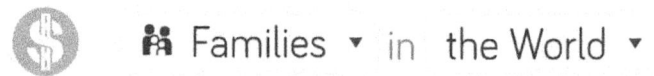

Figure 19: Family Variable

Selecting this variable will show images of all the different families participating in Dollar Street (see Figure 20). Students can be prompted to ignore the text on the images (which displays the location of the photo and the monthly consumption value in dollars) and just focus on the images themselves. Students can ask:

What does the typical family photograph look like for the Dollar Street families?

Level B students notice that all over the world families posing for the photos are physically close to one another, often hugging. Small children are held in the arms of a family member. How can the body language captured in the photos be described? Is it similar across the world? Families seem at ease – some are smiling, others are not. Students can also notice that family size (number of individuals identified as family) varies.

Figure 20: Families from Dollar Street

These photographs demonstrate the groups labeled as family might differ. Some families include an elder generation (presumably grandparents), others have one parent, and some have two. Some families do not have children, and other families might have children who are grown and not pictured.

Despite the differences in appearance or age, the most striking observation is that the concept of family as a group together seems to be the same across all the countries represented. Students can be encouraged to click on a picture to learn more about a specific family.

A second variable Level B students can explore is Front Doors of the families' living spaces (see Figure 21).

Figure 21: Front door variable

Students in Level B can look at the images of the front doors of the families' homes to develop a description of a door, including typical materials, operation, and function. In general, the photos of doors show some type of cover over an opening. At its most basic level, a door is the same all over the world. The doors also seem to be made from similar materials— mostly wood, though sometimes metal or cloth. Figure 22 shows four doors across the economic spectrum that all are made of wood:

Figure 22: Doors made of wood

There appears to be some variation in doors when considering the entire monthly consumption range. In the low consumption group, for example, the doors often do not have locks and are less ornate, while in the high consumption group, the doors all have locks and sometimes many locks.

Figure 23: Selection Bar

To explore the association between consumption level and door types, students may categorize consumption amounts into either high, medium, or low consumption. For example, a medium category would include those families in the middle range of the "poorest to richest" scale provided on the site. Students can select only these families by adjusting the selection bar (see Figure 23).

High Consumption:

Medium Consumption:

Low Consumption:

Figure 24: Doors for different consumptions

Figure 24 shows some photo examples of doors in the different categories of high, medium, and low consumption.

Although there are differences among the doors, what is most striking are the similarities. Students may see very little variation among the doors across the world, particularly when focused on a specific economic band.

Level B students can also examine more variables, such as the Stoves variable (see Figure 25).

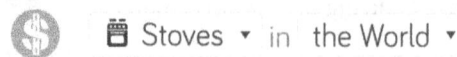

Figure 25: Stoves variable

Students might observe several differences with the Stoves variable. First, stoves in families who consume less per month appear to be very different from those in families who consume more per month (see Figure 26).

Figure 26: Stoves for different consumptions

Families who consume more per month appear to have stoves that are electric or gas (see Figure 27). These stoves might be placed within a countertop or stand alone as a single appliance. In the lower consumption families,

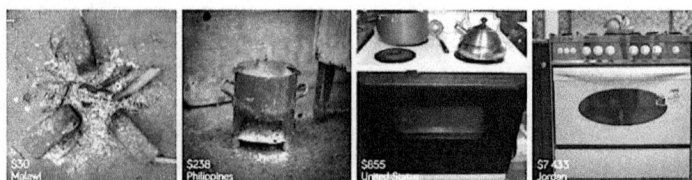

Figure 27: Stoves with different heating method

cooking stoves appear to be often on the ground or directly structured over a flame. Wood burning is common in low-medium consumption families. Any gas-burning stoves in low consumption families often have only a single burner, not multiple burners. Although all the families across the world appear to use heat to cook, their modes of doing so vary.

For more information on the Dollar Street Project as well as some observations about cooking across the world, see the following TED talk by the founder, Anna Rosling Ronnlund, of the project:

https://www.youtube.com/watch?time_continue=3&v=u4L130DkdOw

Interpret results

The statistical investigative question asked students to compare families and family living spaces across the world. Overall, the picture data revealed a similar concept of family across the economic spectrum as well as across the countries of the world. Doors were similar everywhere in the world, though there was some variation across consumption levels. On the other hand, cooking stoves varied across the consumption spectrum. Overall, living spaces appear to vary more economically rather than geographically.

The Dollar Street project is a collection of 300 families at this point. None of these conclusions should be interpreted in a generalizable way. Instead, this is an observational study of data presented as pictures. Students in Level B must understand that conclusions drawn from this picture data are limited in scope due to the limitations of the data. The Dollar Street example brings awareness to students of how photographic data from social media sites can easily shape and shift perceptions or conclusions that are far beyond the scope of the data. The development of healthy skepticism in terms of statistics requires students to ask hard questions of data and those collecting, storing, and/or presenting it.

Students in Level B must be exposed to unconventional data to broaden their conceptualization of the impact of data and statistics in our world. Hopefully, by using data sets such as Dollar Street, students can be pushed to think critically about the perceptions they may have about how humans live across the world and about the importance of further research on the topic.

Example 9: Memory and music - Comparative experiments

Another important study design appropriate for Level B students to explore is the comparative experimental design. Comparative experimental studies involve comparisons of the effects of two or more treatments (experimental conditions) on some response variables. At Level B, studies comparing two treatments are adequate. This type of analysis is common for students participating in science fair competitions.

Formulate statistical investigative questions

It is common for students to listen to music while studying. Some students claim that listening to music helps them focus and puts them in a positive mood. To investigate this topic, Level B students may design a comparative experimental study to explore the effects of listening to music on one's ability to memorize words. A statistical investigative question may be:

Are students able to memorize more words while listening to music than not listening to music?

Collect data/consider data

The class needs to develop a design strategy for collecting experimental data to answer the investigative question. This will involve the students identifying and, as much as possible, controlling for potential extraneous sources of variability that may interfere with interpreting the results.

One simple experiment to conduct with a class would be to randomly divide the class into two equal-sized groups. Students develop the understanding that random assignment is an important part of experimental design because it tends to average out differences in student ability and other characteristics that might affect the response. Random assignment also allows for causal statements to be made.

Suppose that 28 students are randomly assigned into two groups of 14. When possible, the class should use technology for the random group assignment, but tools such as a deck of cards would also work. For example, a teacher can shuffle 28 cards, 14 red cards and 14 black cards, and hand one out to each student. Those who receive a red card will participate in the group that listens to music, and those who receive a black card will participate in the silence group.

The class should develop an experimental procedure and agree on its details. For example, each participant may have two minutes to study a list of 20 words, followed by a one-minute pause, and then have two minutes to write down as many of those words as possible. Participants in the music group will listen throughout the experiment to a particular song with lyrics, while the control group remains in silence the entire time. The number of words correctly remembered is the response variable of interest. The data set thus includes two variables: the participant's randomly assigned group, which is a categorical variable, and the number of correctly remembered words, which is a discrete quantitative variable. Students can use technology to record student data.

Analyze data

Students should calculate summary statistics for the music-listening group and for the silence group. Table 7 shows example data. At some point, students should manually calculate these numbers from a small data set to understand their composition. However, with larger data sets and when collecting data with technology, it is important to know how to use technology to calculate the five-number summary.

Table 7: Five-Number Summaries of number of words memorized in music and silence groups

	Number of words memorized	
	Music Group	Silence Group
Minimum	3	6
First Quartile	6	8
Median	7	10
Third Quartile	9	12
Maximum	15	14

Boxplots that use a common scale for each group are a good data visualization (See Figure 28).

Interpret results

These results suggest that the students participating in the study generally memorize fewer words when listening to music than in silence. With the exception of the maximum value in the music group, all of the five-number summary measures for the music group (M) are lower than the corresponding summary

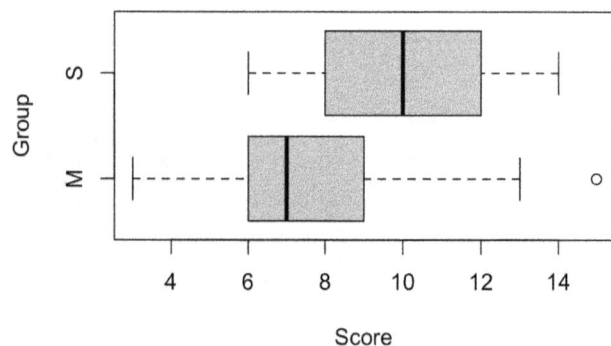

Figure 28: Comparative boxplots for memory data – Music (M) and Silence (S)

measures for the silence group (S). The variation in the middle-50% of scores is similar for both groups. Distribution S appears to be roughly symmetric, while distribution M is slightly right-skewed. Considering the variation in the scores and the separation in the boxplots (medians are outside the overlap of the central boxes), a difference of 3 between the medians is quite large.

The students may wonder if the type of music was an important factor in the outcome. For example, did the presence of words within the song contribute to the variation between boxplots? They may consider repeating the experiment with a new group of students, this time using only instrumental music. Level B students should understand the scope of inference of this experiment. Students were not randomly selected, but they were randomly assigned to the treatments. The lack of random sampling limits the scope of inference to the class rather than to the general population. The inclusion of random assignment allows for causal statements with respect to the treatments for the students in the class. Level C students will further develop the role of randomization in statistics.

Summary of Level B

Students who understand the statistical concepts of Level B can begin to appreciate that statistical reasoning is a problem-solving process. They will get experience formulating their own statistical investigative questions; collecting and considering appropriate data through various sources; analyzing data through graphs and summary measures; and interpreting results with an eye toward inferences of causation or inferences from a sample to a population. As they begin to formulate their own statistical investigative questions, students become aware that the world around them is filled with data that affect their lives, and they begin to appreciate that statistics can help them make decisions based on data. With their understanding of the statistical problem-solving process, Level B students can be encouraged to examine claims in the media such as those around health decisions (vaccines, flu shots, etc.), environmental concerns (sea level, CO_2 levels, etc.), or educational policy (curriculum, assessment, etc.). See www.nctm.org/gaise for further examples.

As students become comfortable with Level B concepts, they will be prepared for the further depth of Level C.

Level C

Introduction

The Role of Technology

The Role of Probability in Statistics

Essentials for Each Component

Example 1: Level B revisited: Darwin's Finches – MAD to Standard Deviation

Example 2: Level A and B revisited: Choosing music for the school dance (continued) – Generalizing Findings

Example 3: Choosing music for the school dance (continued) – Inference about Association

Example 4: Effects of Light on the Growth of Radish Seedlings – Experiments

Example 5: Considering Measurements when Designing Clothing – Linear Regression

Example 6: Napping and Heart Attacks – Inferring Association from an Observational Study

Example 7: Working-age Population – Working with Secondary Data

Example 8: Classifying Lizards – Predicting a Categorical Variable

Summary of Level C

Introduction

Guidelines at Level C build on the foundation for statistical reasoning developed at Levels A and B. Illustrative activities at Level C should continue to emphasize the four components in the statistical problem-solving process and solidify the spirit of genuine statistical practice.

At Level C, students understand how to use questioning throughout the statistical problem-solving process and how to use information from data to answer appropriately framed statistical investigative questions. They recognize non-traditional data (text, sounds, pictures) alongside traditional data (numbers with context). They choose and use appropriate data analysis tools (graphical displays, tabular displays, and numerical summaries). Level C students also engage in multivariable thinking, and the types of statistical investigative questions expand to include questions concerning causality and prediction.

At Levels A and B, students collected data from whole groups, samples, and simple experiments. At Level C, these types of studies are emphasized at a deeper level. Students strengthen their understanding of the role of random sampling and random assignment. Statistical studies at Level C develop a more formal understanding of inferential reasoning and its use of probability. Students also focus on explaining statistical reasoning to others.

Many data that students encounter in their daily lives come not from formal studies, but from sensors, "bots" (e.g., the number of times one clicks on a particular advertisement or the number of cars that cross a particular intersection) or interactions with technology (e.g., devices tracking the amount of screen time one is exposed to in a week). Sometimes data may represent the entire population, such as data about music choices from all users of a music streaming service. Classes of students might collect their own data using their mobile devices to document the incidence of graffiti in their town or to identify dangerous crosswalks. Level C students should understand that these data do not always support formal statistical inferential reasoning. Nonetheless, these data can tell a partial story that describes potential patterns or processes, or that provides hypotheses for more formal data collection.

At Level C, students develop a more sophisticated notion of data that accounts for errors and missing values. They deepen their analysis by integrating technology into their practice, transforming variables or creating new ones. Students understand that data are often stored as files on a local computer, server, or in the cloud, and that these files can be shared and altered, raising the need for reproducible practices and ethical data handling.

Level C moves students past descriptive statistics of the whole group or population to incorporating notions of chance and probability to draw general inferences and inferential comparisons about two groups, as appropriate. Simulations are employed to enhance probabilistic reasoning.

Level C students understand sample-to-sample variability. They can simulate approximate sampling distributions and use them to compute p-values. The p-value is a commonly used statistic for addressing the question *Can this outcome be due solely to chance?* In recent years, the use and interpretation of the p-value in research findings has been appraised, leading to recommendations for more careful attention to how statistical findings are reported (e.g. Nuzzo, 2014; Wasserstein & Lazar, 2016). In statistics, Level C students should understand the importance of asking and answering the question *Can this outcome be due solely to chance?* Level C students can evaluate a simulated sampling distribution to determine which outcomes are considered extreme and which outcomes are probable.

The Role of Technology

Whenever possible, technology should be used to develop an understanding of statistics and to carry out statistical investigations. Integrating technology into the curriculum can be challenging. Not all classrooms are equipped with the necessary hardware or software, and changes in technology are so frequent that planning can be difficult. Technology can also exacerbate equity concerns because some students have greater access to technology or support for technology at home than other students. Support for technology can vary greatly, even within local regions. Still, modern statistical practice is inseparable from technology, and many software tools and applets are freely available to enhance students' understanding, and so it is recommended that technology be embraced to the greatest extent possible.

Abstract statistical concepts, such as the notion of sample-to-sample variability, can be made more concrete through technology. In real life, rarely are opportunities available to take multiple random samples and directly observe the chance-induced variation of estimates. But simulations and applets can allow students to directly observe the variability from repeated samples, and from this develop a deeper understanding of the theory. Simulations and applets also can illustrate how random assignment is necessary to make causal statements, thus giving students a deeper understanding of the role of chance and probability in statistical inference.

Computational thinking is increasingly emphasized within schools and policy by including and promoting computer science in the curriculum. A statistics class that uses a statistical programming language provides a rich complement to computer science courses. Because of this, the importance of statistical thinking is greater than ever. As data are being used in computer science and other disciplines, it is extremely important that Level C students understand the scope of data, the limitations of data, and the types of questions that can be asked about data. There are also statistical software packages designed specifically for learners of statistics and that can facilitate deeper statistical thinking. Teachers must choose technologies based on local needs and constraints. Different software packages have their own strengths and weaknesses, but the use of any software package is preferable to routinely performing calculations by hand. A thoughtful implementation of technology provides for richer and deeper statistical investigations that will allow students to fully appreciate the importance of statistics. Several websites provide links to open-source collections of tools for learning and doing statistics (e.g., causeweb.org and amstat.org/education); for additional resources, see www. nctm.org/gaise.

The Role of Probability in Statistics

Teachers and students must understand that statistics and probability are not the same. Statistics *uses* probability in a similar way as physics uses calculus. Students of statistics should develop a strong understanding of some central concepts and applications of probability. These conceptual understandings will be more useful than procedural skills to calculate probabilities. When calculations are needed, probabilities will most often be calculated based on given distributions. Students can use technology or tables for these calculations.

Answering the question *Could the outcomes observed be due solely to chance?* requires an understanding of the randomness in the process that generated the observations. *What values can chance produce? How often do those values occur?* For an example of how technology can be used to create representations of probability, see Pfannkuch & Budgett (2017). Designing and running simulations is an essential tool for developing this understanding.

For example, what do scores on a 20-question true/false exam look like if a student chooses answers based solely on chance? Students should recognize that there is great variability in the potential outcomes. Any score from 0 correct to 20 correct is possible. However, not all of these scores are equally likely. Students can simulate multiple test scores by repeatedly flipping a fair coin 20 times and using the number of heads to represent the number of correct answers on the test. They will then see that some test scores occur more often than others. All but the very skeptical would conclude that a student who got 100% on a well-developed test knew the material. On the other hand, a student with a score of 50% performs similarly to a student who is guessing.

A common misconception of probability is that events with low probability never occur and events with high probability always occur. For example, a candidate who is estimated to have an 80% chance of winning an election will not necessarily win. In a hypothetical universe where there are many elections and the candidate has an 80% chance of winning, the candidate will lose in 1 out of 5 (20%) of the elections. Likewise, a forecast of a 5% chance of rain does not mean it will be dry today. It either will or will not rain; the forecast merely states that on days like today, 5% of them have rain.

Consider the question *Is there an association between eating red meat every day and having a heart attack?* Predicting outcomes for any particular individual has high uncertainties. Instead, statisticians answer such questions based on the characteristics of groups. Probability provides a way of describing risks and propensities of very large groups of people—not of individuals within those groups.

We use chance-driven variability in statistical inference to quantify the uncertainty in estimation. For example, if the percentage of people in the United States who wear contact lenses is to be estimated based on a sample of people in our school, Level C students must understand that the percentage in their sample may be wildly different from the true percentage in the population. Yet because their sample is not random, they cannot know how different it may be. If instead people were randomly selected from the population, then the uncertainty can be quantified with a margin of error. For example, they might estimate the percentage of adult contact wearers in the United States to be 12% with a margin of error of plus-or-minus 5 percentage points. A student with a strong conceptual understanding of probability will interpret this to mean that she can be highly confident that the true percentage of contact wearers in the population is between 7% and 17%. But, even so, it is not impossible that the true percentage is outside this interval.

Two important and related concepts from probability are conditional probability and independence. Conditional probability provides a means for updating probabilities when more information is known. For example, one might conclude that there is a 60% chance of a randomly selected student liking rap music. But perhaps there is an association between grade level and music preference. In such a case, conditional probability might be used to explain that if a student in fourth grade is randomly selected, there is a 45% chance they will prefer rap, but if a student in sixth grade is selected, then there is a 70% chance. In other words, conditioning on knowledge of the student's grade level changes the probability that a randomly chosen student will like rap. When probabilities are not affected by conditional knowledge, the variables are independent. If music preference is independent of grade level, then the probability is the same for selecting a rap lover in fourth grade as sixth grade. Here, this probability would be 60%, the probability of randomly selecting a student who likes rap music regardless of grade level.

The notion of independent observations is fundamental to statistics. Essentially, when observations are independent, they provide unique pieces of information to our dataset. For example, suppose that cookies are being weighed for a bake sale to confirm that each cookie weighs about the same. If the fresh cookies melt somewhat on the scale and leave a residue, the weights of cookies weighed later may be higher than those weighed earlier. In probability terms, knowledge of the order in which a cookie is weighed affects the probability that the cookie will be too heavy. In this case, the weights are not independent measurements. If instead the scale is cleaned and recalibrated after each use, the value recorded for one cookie is independent of the value recorded for any other, and independence is achieved.

Modern statistical practice includes prediction in multivariable contexts. In these cases, the focus of the investigation is an individual and not a group. It might be helpful to use a person's dietary history to estimate their chance of a heart attack within the next 10 years. Based on these probabilities, people or objects might be classified into groups (e.g., "will have a heart attack" or "will not have a heart attack"), and future data will be used to evaluate the success rates (or error rates) of these classifications.

Essentials for each component

I. Formulate statistical investigative questions

→ Formulate multivariable statistical investigative questions and determine how data can be collected and analyzed to provide an answer

→ Pose summary, comparative, and association statistical investigative questions for surveys, observational studies, and experiments using primary or secondary data

→ Pose inferential statistical investigative questions regarding causality and prediction

II. Collect/consider data

→ Apply an appropriate data collection plan when collecting primary data for the statistical investigative question of interest

→ Distinguish between surveys, observational studies, and experiments

→ Understand what constitutes good practice in designing a sample survey, an experiment, and an observational study

→ Understand the role of random selection in sample surveys and the effect of sample size on the variability of estimates

→ Understand the role of random assignment in experiments and its implications for cause-and-effect interpretations

→ Understand the issues of bias and confounding variables in observational studies and their implications for interpretation

→ Understand practices for handling data that enhance reproducibility and ensure ethical use, including descriptions of alterations, and an understanding of when data may contain sensitive information

→ Understand how concerns about privacy and human subjects may affect the collection and distribution of data

→ Understand that in some circumstances, the data collected or considered may not generalize to the desired population, or this data may be the entire population

III. Analyze the data

→ Use technology to subset and filter data sets and transform variables, including smoothing for time series data

→ Identify appropriate ways to summarize quantitative or categorical data using tables, graphical displays, and numerical summary statistics, which includes using standard deviation as a measure of variability and a modified boxplot for identifying outliers

→ Summarize and describe relationships among multiple variables

→ Understand how sampling distributions (developed through simulation) are used to describe the sample-to-sample variability of sample statistics

→ Develop simulations to determine approximate sampling distributions and compute p-values from those distributions

→ Describe associations between two categorical variables using such measures as difference in proportions and relative risk

→ Describe the relationship between two quantitative variables by interpreting Pearson's correlation coefficient and a least-squares regression line

→ Use simulations to investigate associations between two categorical variables and to compare groups

→ Construct prediction intervals and confidence intervals to determine plausible values of a predicted observation or a population characteristic.

IV. Interpret results

→ Use statistical evidence from analyses to answer the statistical investigative questions and communicate results through more formal reports and presentations

→ Evaluate and interpret the impact of outliers on the results

→ Understand what it means for an outcome or an estimate of a population characteristic to be plausible or not plausible compared to chance variation

→ Interpret the margin of error associated with an estimate of a population characteristic

→ Acknowledge the presence of missing values and understand how missing values may add bias to an analysis

→ Use multivariate thinking to understand how variables impact one another

→ Communicate statistical reasoning and results to others in a variety of formats (verbal, written, visual)

→ Understand how to interpret simulated p-values appropriately

Example 1: Level B revisited: Darwin's Finches – MAD to Standard Deviation

Formulate statistical investigative questions

Students at Level B explored the secondary data set about finches by considering the statistical investigative question:

Does one of the finch species on the Galapagos Islands, Cactus Finch or Medium Ground Finch, tend to have a greater beak depth than the other?

Level C students can revisit this same statistical investigative question but enhance their analysis from those used at Level B.

Collect/consider data

Level C students can use the same data set as students in Level B with a sample of 101 finches with 59 Medium Ground Finches (MGF/fortis) and 42 Cactus Finches (CF/scandens).

The band represents the ID number of an individual finch. The species is a categorical variable with two categories, CF for scandens or MGF for fortis. The beak length is a continuous quantitative variable representing the measurement of the length from the

Table 1: Summary table of available variables

101 Total Finches	
Variable names	Description
Band	Id# of finch
Species	CF or MGF
Beak length	Length of beak in mm
Beak depth	Depth of beak in mm
Year	Year observation recorded
Drought	Before or after 1977 drought

base of the beak to the tip, measured in millimeters. The beak depth is a continuous quantitative variable representing the height of the widest point in the beak, also measured in millimeters. The year variable identifies the year in which the measurements were taken. Lastly, the drought variable is a categorical variable with two categories indicating whether the measurements were taken before or after 1977, the year in which there was a major drought.

The data collection was designed to have about 30 finches for each species from before the drought and after the drought; however, there were fewer Cactus Finches observations collected after the drought.

Analyze the data

In Level B, students described the distributions of beak depths for both the sample of MGF/fortis and CF/scandens as approximately symmetrical and unimodal. Because the distributions were approximately symmetric with no extreme outliers, the mean was chosen as the measure of center.

Level B students developed the Mean Absolute Deviation (MAD) as a measure of variability from the mean for the distribution of a quantitative

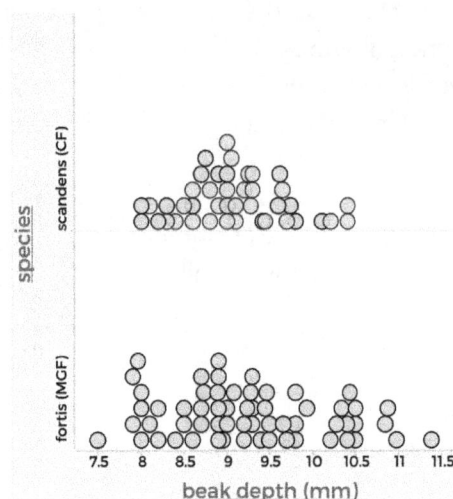

Figure 1: Beak depth for CF/scandens MGF/fortis species

variable. The MAD is the average distance of each data value from the mean. This calculation uses the absolute value of the (observed – mean) to make all the distances positive. That is:

$$MAD = \frac{\text{Total distance from the mean for all values}}{\text{Number of data values}}$$

$$= \frac{\sum |observed - mean|}{n}$$

Students at Level C discover that another way to make the difference (*observation – mean*) positive is by squaring them. If we square the differences, sum them, and divide by *n – 1*, the result is called the variance (V), which can be written as:

$$V = \frac{\sum (observed - mean)^2}{n - 1}$$

The variance is the average squared distance of each value from the mean, and it's useful because it can be added across independent random variables. However, it is not in the same base units as the given data. Taking the square root of the variance, however, results in a quantity in the same base units. This quantity is the standard deviation (SD), which can be written as:

$$SD = \sqrt{\frac{\sum (observed - mean)^2}{n - 1}}$$

Students may notice that our goal was to get the average squared distance from the mean and so then might wonder why we divide by *n – 1* rather than *n*. When a sample from the population is being analyzed instead of the entire population, the denominator is n-1 instead of n. Because the sample mean is used to find the differences in the sample, then n-1 differences determine the last difference as the differences must sum to 0. Typically, the observed sample values fall, on average, closer to the sample mean than the values in the actual population. Thus, if the sum is divided by n, the sample SD will underestimate the population SD. To correct for this mathematically, the sum is divided by n-1.

Table 2: Summary statistics for finches

Species	Mean	MAD	SD
CF	9.07 mm	0.49 mm	0.62 mm
MGF	9.26 mm	0.73 mm	0.90 mm

The calculation for the SD formula, with all its moving parts, is more challenging for students compared to the MAD. What is typically lost by students in this non-intuitive calculation is how to interpret it. The MAD can serve as a bridge or scaffold to the SD, since it is more intuitive mathematically. By learning how to interpret the MAD, students have developed a conceptual understanding that is similar to interpretating the SD.

For the finch data, the MAD of 0.73 mm (see Table 2), can be interpreted as how far the MGF beak depth values lie from the mean of 9.26 mm, on average. In other words, the beak depth is typically 9.26 ± 0.73 mm (between 8.53 mm and 9.99 mm). A similar interpretation can be applied for the standard deviation: the beak depth for MGF typically differs from the mean of 9.26 mm by ±0.90 mm.

The MAD and SD will generally be similar in value. We would expect that the SD for the MGF birds is greater than the SD for the CF birds, since the observations for MGFs have a greater range and are less clustered around the mean than the observations for the CFs. Level C students might notice based on the calculation that the SD is more sensitive to outliers. The MAD is a more robust statistic since it is less sensitive to outliers, and it may be more useful in some situations.

Why then is the SD used more in practice than the MAD? A major reason is the SD's mathematical relationship with bell-shaped, symmetric, unimodal distributions. A special distribution of this type, the Normal Distribution, has the following characteristics described by the Empirical Rule (see Figure 2):

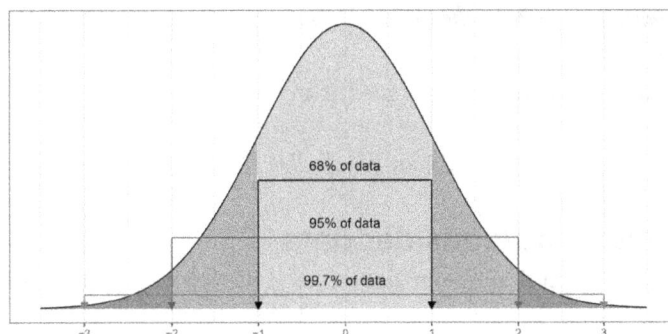

Figure 2: Normal Distribution and the Empirical Rule (This figure was originally published as Figure 14.3 in *Focus on Statistics: Investigations for the Integration of Statistics into Grades 9-12 Mathematics Classrooms.*)

- roughly 68% of the observations are expected to fall within 1 SD of the mean

- roughly 95% of the observations are expected to fall within 2 SD of the mean

- All or nearly all of the observations are expected to fall within 3 SD of the mean

Interpret the results

As in Level B, students at Level C would conclude there is no meaningful difference between beak depth for MGF and CF finches. The distributions have similar centers and variability as measured by the mean and standard deviations. The distributions have similar shapes and mostly overlap with very little separation.

Example 2: Level A and B revisited: Choosing music for the school dance (continued) – Generalizing Findings

Formulate statistical investigative question

A survey of student type of music preferences was introduced at Level A, where the analysis consisted of making counts of student responses and displaying the data in a bar graph. At Level B, the analysis was expanded to consider relative frequencies of preferences and cross-classified responses for two types of music displayed in a two-way table.

In previous examples, the data were selected from a class (or grade level), and generalizations did not formally extend beyond that class or grade level. In Level C, students consider how to generalize findings from a sample of a few students within a school to the entire school.

A student believes that all of the music at the school dance should be rap because "most students like rap." To test this claim, this student poses the following statistical investigative question:

Do most students like rap?

In other words:

Do more than 50% of the students at our school like rap?

This question could also be stated as a claim:

More than 50% of the students at our school like rap.

Collect data/consider data

Recall the Level B questionnaire included the following survey questions:

Q1. Check yes for any of the following music types you like. Check no for any you don't like.

	Yes	No
Rap	☐	☐
Rock	☐	☐
Country	☐	☐
R&B	☐	☐
Pop	☐	☐
Classical	☐	☐
Alternative	☐	☐

Q2. What is your favorite type of music?

- ☐ Rap
- ☐ Rock
- ☐ Country
- ☐ R&B
- ☐ Pop
- ☐ Classical
- ☐ Alternative
- ☐ Other

To generalize to all students at a school, a representative sample of students from the school is needed. At Level C, a simple random sample of 60 students from the school was selected to be surveyed. The results can then be generalized to the school (but not beyond), and the Level C discussion can center on basic principles of generalization—that is, statistical inference.

Analyze the data

The statistical investigative question involves only the "Like Rap?" variable. Because 58% of the students in the sample liked rap music (which is more than 50%), there is some evidence to support the statement that more than 50% of the students at the entire school like rap. However, it is possible that the observed value of 58% resulted from chance due to random sampling, and the true percentage of students who like rap is 50% or less (in which case, the student's claim is not correct).

We can use a simulation to examine whether or not an observed proportion of 58% successes (liking rap) is possible (or probable) when the population has only 50% successes. We can set up a hypothetical population that has 50% successes. This can be modeled using coin flips. Students can simulate this by tossing a coin 60 times. As a class, they can combine their results to form a distribution of simulated sample proportions to investigate how many of them saw 58% or more heads. They should discuss whether this leads them to believe that a result of 58% "successes" in a sample is a rare outcome or a common outcome when the true percentage is 50%. Unless the class is very, very large, they will not have too many simulated observations to draw a strong conclusion. An applet or statistical software will allow them to produce many more. By flipping a virtual coin 60 times, recording the number of successes, and repeating many times, an approximate sampling distribution similar to Figure 3 can be generated.

Interpret the Results

Based on this distribution of simulated sample proportions, a sample proportion greater than or equal to 0.58 occurred 17 times out of 100 (counting the number of dots falling at or to the right of 0.58 on the dotplot) just by chance variation alone when the actual population pro-

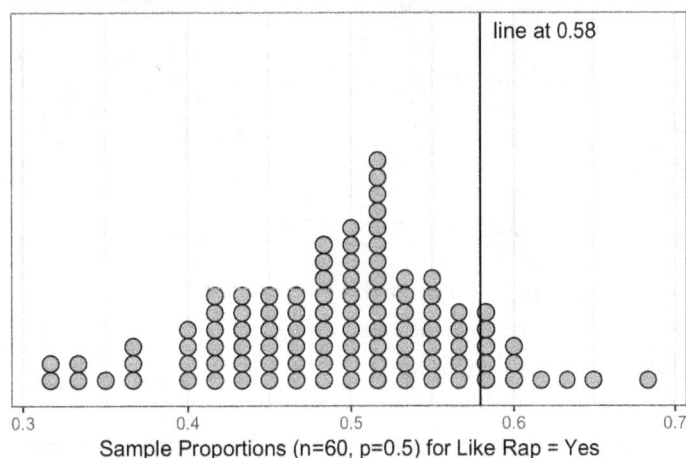

100 Simulations

line at 0.58

Sample Proportions (n=60, p=0.5) for Like Rap = Yes

Figure 3: Distribution of simulated sample proportions where the population proportion = 0.50

portion is 0.50. This suggests the result of 0.58 is not a very unusual occurrence when sampling from a population with 0.50 as the true proportion of students who like rap music. So, the evidence in support of the claim that more than 50% of students at the school like rap is not very strong. Presuming the value of 50% is, in fact, correct, the fraction of times the observed result is matched or exceeded in repeated repetitions of the experiment (0.17 in this investigation) is an approximation of what is called the p-value. (Using more than 100 repetitions of the simulation would provide a better approximation.) A small p-value would have indicated that if the population proportion was 0.50, it is very unlikely that a sample proportion of 0.58 (where the vertical line is placed in the graph) would have been observed.

By way of contrast, another student might pose the following statistical investigative question:

Do more than 40% of the students at our school like rap?

This leads to the claim: More than 40% of the students at our school like rap.

To investigate this student's question, samples of size 60 must now be repeatedly selected from a hypothetical population that has 40% successes. Then, the observed proportion of 0.58 can be compared to the simulated proportions drawn from a population where the proportion is 40%. Figure 4 shows the results of 100 simulated samples. The observed result includes one sample that produced a proportion greater than 0.58. Thus, the approximate *p*-value is 0.01, which is very small, indicating it is not likely that a population in which 40% of the students like rap music would have produced a sample proportion of 0.58 or greater in a random sample of size 60. This small *p*-value provides very strong evidence in support of the student's claim that more than 40% of the students in the entire school like rap music.

Another way of stating the above is that 0.5 is a plausible value for the true population proportion (*p*), based on the sample evidence, but 0.4 is not. A confidence interval can be used to describe a set of plausible values using the margin of error. The margin of error for a 95% confidence level for a sample proportion is:

$$2 * \sqrt{\frac{p(1-p)}{n}}$$

The formula can be motivated through students' repeated exposure to simulations in which they draw many random samples from a population with a known value of p and examine the resulting empirical distribution of sample proportions. Further investigation will show students

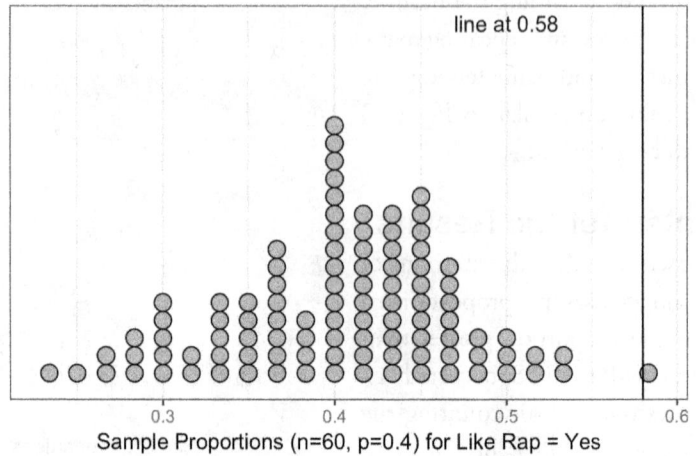

100 Simulations

line at 0.58

Sample Proportions (n=60, p=0.4) for Like Rap = Yes

Figure 4: Distribution of simulated sample proportions where the population proportion = 0.40

that approximately 95% of the sample proportions in the simulated distribution shown in Figure 4 (where the population proportion is 0.4) lie within a distance of

$$2 * \sqrt{\frac{(0.4)(0.6)}{60}} \approx 0.126$$

of the true value of p.

However, in most situations, the true value of p is not known, and so we estimate it with the sample proportion $\hat{p}$.

$$2 * \sqrt{\frac{\hat{p}(1-\hat{p})}{n}}$$

In our study of type of musical preferences the true proportion of students who like rap is unknown. Our sample proportion ($\hat{p}$= 0.58) is our "best estimate" for what the value of might be, so the margin of error can be estimated to be:

$$2 * \sqrt{\frac{\hat{p}(1-\hat{p})}{n}} = 2 * \sqrt{\frac{(0.58)(0.42)}{60}} \approx 0.127$$

Thus, any proportion between 0.58 – 0.127 = 0.453 and 0.58 + 0.127 = 0.707 can be considered a plausible value for the true proportion of students at the school who like rap music. Notice that 0.5 is well within this interval, but 0.4 is not. This is consistent with our previous conclusion: 0.5 is a plausible value for the true population proportion, but 0.4 is not.

Confidence intervals and margins of error are commonly reported in the media when reporting results of polls or other studies. In the context of education, student test scores are also commonly reported with intervals and ± margin of error. It is important for students at Level C to understand how to interpret confidence intervals as a part of statistical literacy.

Example 3: Choosing music for the school dance (continued) – Inference about Association

Formulate statistical investigative questions

Another type of investigative question that could be asked about the students' music preferences is:

Do those who like rock music tend to also like rap music more than those who do not like rock music?

In other words, *is there an association between liking rock music and liking rap music?*

Collect data/consider data

The same data from the random sample of 60 students can be used to answer this question. Students should recognize that the relevant variables in this collection are Likes Rap and Likes Rock.

Analyze the data

For the students in this class, the association between liking rock and liking rap music can be summarized in a two-way table as in Table 3.

According to Table 3, a total of 37 students in the survey like rock music. Among those students, the proportion who also like rap music is 30/37 = 0.81. Among the 23 students who do not like rock music, 5/23 = 0.22 is the proportion who like rap music. The difference between these two sample proportions (0.59) suggests there may be a strong association between liking rock music and

Table 3: Two-Way Frequency Table

		Rock		
		Yes	No	Total
Rap	Yes	30	5	35
	No	7	18	25
	Total	37	23	60

liking rap music. But could the relatively large difference in proportions observed in the sample simply be due to chance (that is, a consequence only of the random sampling)?

To address this question, students in Level C should consider a hypothetical population in which there is no association between the two variables. In such a population, the proportion of students who like rap would be the same for students who like rock and those who do not like rock. Then it is expected that the proportion who like rap among the 37 students who like rock will be close to the proportion of students who like rap among the 23 students who don't like rock. Essentially, if there is no association in the population, the difference between these two sample proportions is expected to be approximately 0.

To simulate this situation, students can create 60 cards to represent each observation in the sample. Label 35 of these cards with "Likes Rap" and the remaining 25 "Doesn't Like Rap". Students randomize their cards by shuffling well and dealing 37 cards into one pile and the rest into another. The first pile represents those students who like rock, and the second pile represents those who do not like rock.

This round of shuffling and dealing simulates a situation in which there is no association between liking rap and liking rock. Whether or not a "Likes Rap" card ends up in the Likes Rock pile or the Doesn't Like Rock pile is determined completely by chance. The difference in proportions of the "Likes Rap" cards in the two piles is a simulated difference of proportions in a population in which the two variables are not associated. By repeating this process of shuffling and dealing many times and recording the difference in proportions of "Likes Rap" cards in the two piles, students can grow an approximate sampling distribution.

The shuffling and dealing process is slow and cumbersome but important as an initial way to simulate and allow students to experience and visualize the role of randomness in 'chance variation alone'. Once students have developed a comfort level with the simulation process, technology can be used to quickly repeat this process 100 times. The resulting approximate sampling distribution will look similar to Figure 5.

Interpret the Results

The observed difference in proportions from the sample data, 0.59, was never reached in 100 trials. The estimated p-value is therefore $0/100 = 0$, which strongly indicates that the observed difference cannot reasonably be attributed to chance alone. Thus, there is convincing evidence of a real association between liking rock music and liking rap music among all students in the school.

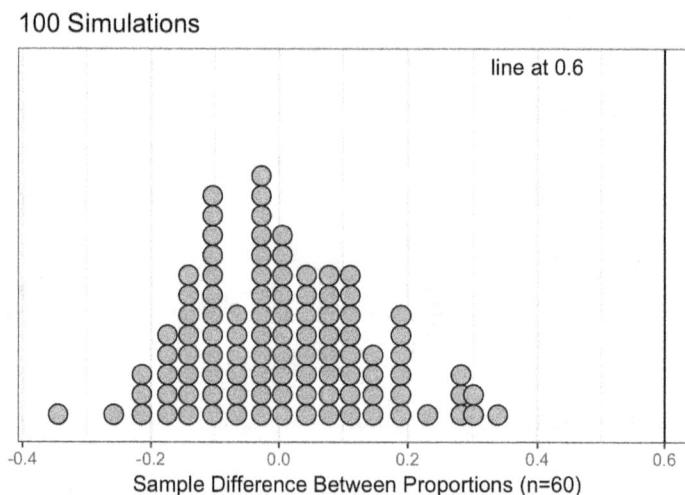

Figure 5: Dotplot showing simulated sampling distribution of difference between two proportions

Example 4: Effects of Light on the Growth of Radish Seedlings – Experiments

Formulate statistical investigative questions

The following statistical investigative question was posed by a class of biology students:

What is the effect of different durations of light and dark on the growth of radish seedlings?

These students then set about designing and carrying out an experiment to investigate the question.

Collect/Consider Data

All possible relative durations of light to dark cannot be investigated in one experiment, so Level C students might decide to focus the experiment on three treatments: 24 hours of light (light), 12 hours of light and 12 hours of darkness (mixed), and 24 hours of darkness (dark). This covers the extreme cases and one in the middle.

With the help of a teacher, the class decided to use plastic bags as growth chambers. The plastic bags permit the students to observe and measure the germination of the seeds without disturbing them. Two layers of moist paper towels were put into a clear plastic bag, with a line stapled about 1/3 of the way from the bottom of the bag (see Figure 6) to hold the paper towel in place and to provide a seam to hold the radish seeds.

One hundred and twenty seeds were available for the study. The size of the growth chambers allowed for only 30 seeds. The class decided to make use of the extra seeds and create four growth chambers, with one designated for the light treatment, one for the mixed treatment, and two for the dark treatment. Thirty of the seeds were chosen at random and placed along the stapled seam of light treatment bag. Thirty seeds were then chosen at random from the remaining 90 seeds and placed in the mixed treatment bag. Finally, 30 of the remaining 60 seeds were chosen at random and placed in one of the dark treatment bags. The final 30 seeds

Figure 6: Seed experiment growth bag for 30 seedlings

were placed in the other dark treatment bag. The students were careful to make the growth chambers as close to identical as possible. With two growth chambers for the same condition, these results could be compared to ensure similar handling. After three days, the lengths of radish seedlings for the germinating seeds were measured and recorded in millimeters.

The data were initially recorded in a summary format similar to that shown in Table 4, which shows the sorted values. There were seeds in each of the treatment types that did not germinate, and these were recorded with an "x" and are considered missing values. Thus, there are a total of 114 observations (28 for Light, 28 for Mixed and 58 for Dark). Level C students should be encouraged to discuss whether omitting the seeds that did not germinate could add bias to the conclusions. Here it was decided that because roughly the same number of seeds failed to germinate in each category, the missing values likely happened "by chance," and thus

Table 4: Radish Seedling Lengths After 3 Days (sorted)

Treatment Type	Seedling Lengths (mm)
1-Light	x,x,2,3,5,5,5,5,5,7,7,8,8,8,9,10,10,10,10, 10,10,10,14,15,15,20,21,21
2-Mixed	x,x,3,4,5,9,10,10,10,10,10,11,13,15,15,15,17,20,20,20,20,20,21, 21,22,22,23,25,25,27
3-Dark	x, 5, 8, 8, 10, 10, 14, 15, 15, 15, 20, 20, 20, 20,20,22,25,25,25,25, 26,30,30,35,35,35,35,36,37,38
3-Dark	x,5,8,8,10,10,10,11,14,15,15,15,16,20,20,20,20,24,25,29,30, 30,30,30,31,33,35,35,40

would not affect the conclusions. Alternatively, if all of the missing values were in one category, this would suggest that the conditions of that category discouraged growth and so missing data were not happening "by chance."

The organization in Table 4 was most convenient for recording their observations. However, for analysis purposes the data could also be represented in long format with each observation (seed) on a separate row, and each variable in a separate column. Table 5 shows the first and last 3 rows of the long format data.

Table 5: Long Format Listing of Radish Seedling Lengths.

Seed #	Growth Bag	Treatment	Length (mm)
1	1	1-Light	x
2	1	1-Light	x
3	1	1-Light	2
...	...	...	...
118	4	3-Dark	35
119	3	3-Dark	35
120	4	3-Dark	40

The observed units are individual seeds and the growth bag is a categorical variable indicating which bag the seed is in (1, 2, 3, or 4), treatment is another categorical variable indicating which treatment the seeds are receiving (1-Light, 2-Mixed, or 3-Dark), and length is a quantitative variable measuring the length of the seeds in millimeters.

The statistical investigation question posed is one of causality; the implicit assumption is that if differences do exist, they are caused by the levels of light. The experimental design, because it employs random assignment to treatment levels, allows concluding that differences, if not caused by chance, are caused by the light levels. But the first step is to rule out the possibility that differences were caused by chance alone.

Analyze the Data

As developed in Levels A and B, interrogating the data is important, and a good first step in the analysis of quantitative data such as these is to make graphs. Boxplots are ideal for comparing the data for the response variable 'length' from more than one treatment, as seen in Figure 7. Both the centers and the variability of the distributions increase as the amount of darkness increases. There are three potential outliers (one at 20 mm and two at 21 mm) in the Treatment 1-Light data.

The summary statistics for these data are shown in Table 6.

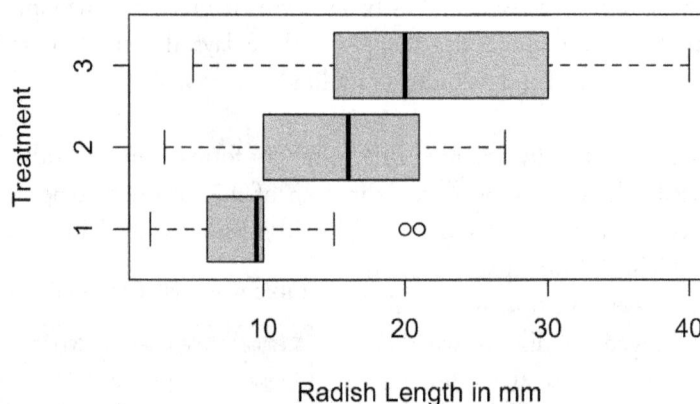

Figure 7: Boxplot showing growth under different conditions 1-Light, 2-Mixed, and 3-Dark

Experiments are designed to compare treatment effects. The original question on the effect of different periods of light and dark on the growth of radish seedlings might be turned into comparison analysis questions about treatment means. Two such questions might be:

Table 6: Treatment Summary Statistics

Treatment	n	Mean	Median	StdDev.
1-Light	28	9.64	9.50	5.03
2-Mixed	28	15.82	16.00	6.76
3-Dark	58	21.86	20.00	9.75

Is there evidence that the 12 hours of light and 12 hours of dark (Treatment 2) group has a significantly higher mean length than the 24 hours of light (Treatment 1) group?

Is there evidence that the 24 hours of dark (Treatment 3) group has a significantly higher mean than the 12 hours of light and 12 hours of dark (Treatment 2) group?

Based on the boxplots and the summary statistics, it appears that the group means differ. This prompts the following analysis question:

Are these differences of means large enough to rule out their occurrence by chance as a possible explanation for the observed difference?

The mean length of seeds in the Treatment 2-Mixed group is 6.2 mm larger than those in the Treatment 1-Light group. Although there is a 6.2 mm difference, it might not be large enough to rule out chance, and so we might not be able to claim a treatment effect. This observed difference might have been due to one bag simply being lucky enough to get a large number of good seeds. It could also be due to the water or light not being equally distributed among the treament groups. But if a difference this large (6.2 mm) is likely to be the result of the randomization of the seeds to the treatments alone, then we should see differences of this magnitude quite often if we were to re-randomize the measurements repeatedly and calculate a new difference in observed means each time. Students can simulate this re-randomization by writing the lengths of all 56 seeds on cards, shuffling the cards, and dealing them into two piles of 28 cards each. The first pile represents the 1-Light treatment group and the second pile represents the 2-Mixed group. Next, students can compute the difference in means between the two piles. This simulated difference is due entirely to chance. By repeating this process many times, students can build an understanding of how the difference of means varies when the treatment has no effect on the growth of the seedlings.

Doing this simulation even once is tedious and slow, but it will be helpful for students when implementing the simulation using technology. Figure 8 was produced by using technology to mix the growth measurements from Treatments 1-Light and 2-Mixed together and to randomly split the measurement into two groups of 28. The difference in means was recorded, and

200 Simulations for Treatment 1-Light and 2-Mixed

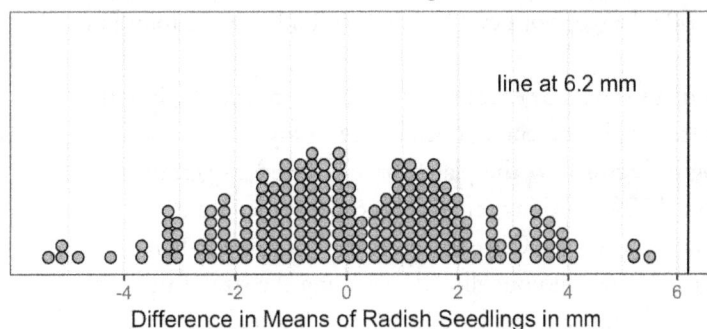

Figure 8: Re-randomization distribution for the difference in means of Treatment 1-Light and Treatment 2-Mixed.

the process repeated 200 times.

The observed difference of 6.2 mm was never exceeded in this simulation of 200 re-randomizations, for an approximate *p*-value of 0/200. This gives strong evidence against the hypothesis that the difference between means for Treatments 1 and 2 is due to chance alone.

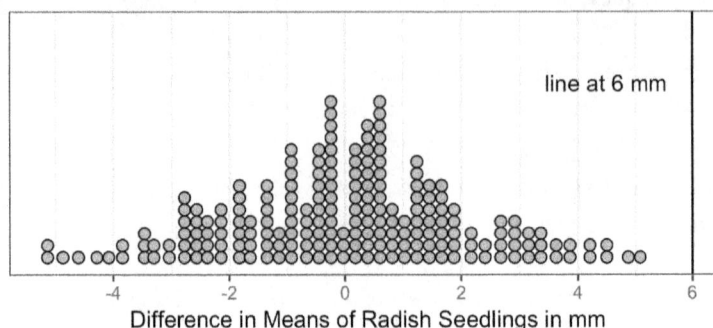

Figure 9: Re-randomization distribution for the difference in means of Treatment 2-Mixed and Treatment 3-Dark

In a comparison of the means for Treatments 2 and 3, the same procedure is used: mixing the growth measurements from Treatments 2-Mixed and 3-Dark together, randomly splitting them into two groups of 28 and 58 measurements, recording the difference in means for the two groups and repeating the process 200 times. The observed difference of 6 mm was never exceeded in 200 trials (see Figure 9), giving strong evidence that the observed difference between the means for Treatments 2 and 3 is not due to chance alone.

Interpret the Results

In summary, two pairs of comparisons of the three treatment groups show differences in mean growth that cannot reasonably be explained by the random assignment of seeds to the bags. This provides convincing evidence of a treatment effect—the more hours of darkness, the greater the growth of the seedling, at least for these three periods of light versus darkness. The randomized assignment allows students to conclude that the differences in growth were caused by the different levels of light.

This analysis is an example of multiple comparisons. To conclude that more hours of darkness causes greater growth, two statistical investigative questions were posed and answered. However, a student in Level C should note that the analyses conducted could be problematic. Only two pairwise comparisons (Treatments 1 and 2 were compared; treatments 2 and 3 were compared) was done while there were three different treatment condition pairs. To fully answer the investigative questions, students should be encouraged to explore the remaining pairwise comparison. Students beyond Level C can also explore potential pitfalls of multiple comparisons with a pairwise approach. More advanced methods exist but are outside the scope of Level C. Such methods may be found in college-level courses.

Students should be encouraged to delve more deeply into the interpretation, relating it to what is known about the phenomenon or issue under study. *Why do the seedlings grow faster in the dark?* A biology teacher might discuss how the plants have adapted to germinate into the light (above ground) as quickly as possible. The seedling cannot photosynthesize in the dark and is using up the energy stored in the seed to power the growth. Once the seedling is exposed to light, it shifts its energy away from growing in length to producing chlorophyll and increasing the size of its leaves. These changes allow the plant to become self-sufficient and begin producing its own food. Even though the growth in length of the stem slows, the growth in diameter of the stem increases and the size of the leaves increases. Seedlings that continue to grow in the dark are spindly and yellow, with small yellow leaves. Seedlings grown in the light are a rich, green color with large, thick leaves and short stems.

Example 5: Considering Measurements when Designing Clothing – Linear Regression

Formulate statistical investigative questions

Clothing designers must create patterns for clothing that are likely to fit their buyers. For clothing to fit people appropriately, a designer must understand the relationship between the dimensions of various physical features. For example, a shirt designer must take into consideration the length of a person's arm in relation to the length of their torso. Suppose a statistics class was tasked with creating a pattern for the costumes for the chorus of the school play. The costumes consist of cloaks that run all the way to the ground, so an actor's feet would not be showing. The cloaks are designed so that the forearm sleeve fits tightly around the arm, running from the elbow to the hand. They ask the following statistical investigative question:

How does a person's forearm length relate to their height? Can forearm length be used to predict height?

Collect/Consider Data

Students in the statistics class decide to use their own body measurements to help them make the cloak pattern. They take their own measurements at home to avoid any potential sensitivity issues in class. Level C students should recognize that taking measurements from students in their class might not be generalizable to the school, however, it offers a basis for their pattern cutting. An important consideration here is to agree on the definition of "forearm" before beginning to take measurements. The cloak forearm is determined to be the measurement from the crease of the elbow on the inside of the arm to the end of the wrist. The data obtained by the students (in centimeters) are provided in Table 7. Each row represents a student.

Students are sometimes unfamiliar with metric units and might want to examine the data in more familiar units. To do this, the data can be transformed. Using dimensional analysis, new variables can be created

Table 7: Heights vs. Forearm Lengths

Forearm (cm)	Height (cm)	Forearm (cm)	Height (cm)
45.0	180.0	41.0	163.0
44.5	173.2	39.5	155.0
39.5	155.0	43.5	166.0
43.9	168.0	41.0	158.0
47.0	170.0	42.0	165.0
49.1	185.2	45.5	167.0
48.0	181.1	46.0	162.0
47.9	181.9	42.0	161.0
40.6	156.8	46.0	181.0
45.5	171.0	45.6	156.0
46.5	175.5	43.9	172.0
43.0	158.5	44.1	167.0

that transform the units from cm to inches, by multiplying the "old" values by 0.393701 in/cm (or dividing by 2.54 cm/in).

Analyze the Data

A good first step in any analysis is to plot the data (see Figure 10). The scatterplot indicates that a linear trend would be a reasonable model for summarizing the relationship between height and forearm length. In Level B, the Quadrant Count Ration (QCR) was considered as a numerical summary for quantifying the strength of the relationship between two quantitative variables. The QCR was based on a count of points in each of the four quadrants of the scatterplot defined by the mean lines for each of the two variables. In Level C, the analysis evolves to using Pearson's correlation coefficient (r) that takes into account the distance of the points from the mean lines. The mean lines are a vertical and a horizontal line placed at the mean of the forearm measurements and the mean of the height measurements, respectively.

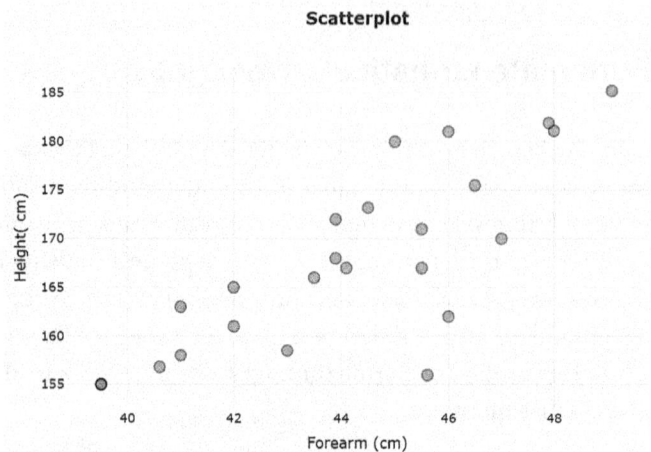

Figure 10: Scatterplot with no regression line

Similar to the QCR, Pearson's *r* is unitless and is always between –1 and +1, inclusive. See Kader and Franklin (2008) for a more detailed discussion. Pearson's r for these data obtained from technology is 0.8. This indicates that there is a strong positive linear association between height and forearm length.

If the "cloud" of points in a scatterplot has a linear shape (a narrow oval or sausage-like pattern in contrast to a more circular pattern), a straight line may be a realistic model of the relationship between the variables under study. Level C students should be introduced to the least-squares regression line (referred to throughout this example as the least-squares line), which is the line that minimizes the sum of the squared residuals. Residuals are defined to be the deviations in the *y*-direction between the points in the scatterplot and the least-squares line. Applets can be used to show the concept of the least-squares line and residuals. The least-squares line also runs through the point where the mean lines cross at (mean x, mean y) of the cloud of points.

The equation for the least-squares line for this relationship (calculated with statistical software) shown in the scatterplot in Figure 11 is:

Predicted Height = 45.8 + 2.76(Forearm Length).

The plot below the scatterplot shows the residuals (Figure 11).

The residuals show variation around the least-squares line, as measured by the standard deviation of the residuals.

Students in Level C should evaluate whether the linear model is appropriate and a good fit. The residual plot reveals that there are no patterns: there are similar amounts of negative and positive residuals and they are both negative and positive across the values of x. This supports the initial impression of the scatterplot that the linear model is appropriate. The correlation coefficient was already noted above as being high (0.8), indicating the linear relationship is strong.

The slope (about 2.8) of the least-squares line can be interpreted as an estimate of the average difference in heights for two persons whose forearms are 1 cm different in length. That is, if one actor's forearm length is 1 cm greater than another, the actor is expected to need a cloak that is about 2.8 cm longer. The intercept of 45.8 centimeters can theoretically be interpreted as the expected height of a person with a forearm zero centimeters long. Yet this is clearly unreasonable, so in this context interpreting the y-intercept of the least-squares line is not sensible. However, the least-squares line can reasonably be used to predict the height of a person for whom the forearm length is known, as long as the known forearm length is in the range of the data used to develop the least-squares line (39 to 50 cm for these data). For example, the predicted height of someone with a forearm length of 42 cm would be:

$$\text{Predicted Height} = 45.8 + 2.76(42) = 161.7\text{cm}$$

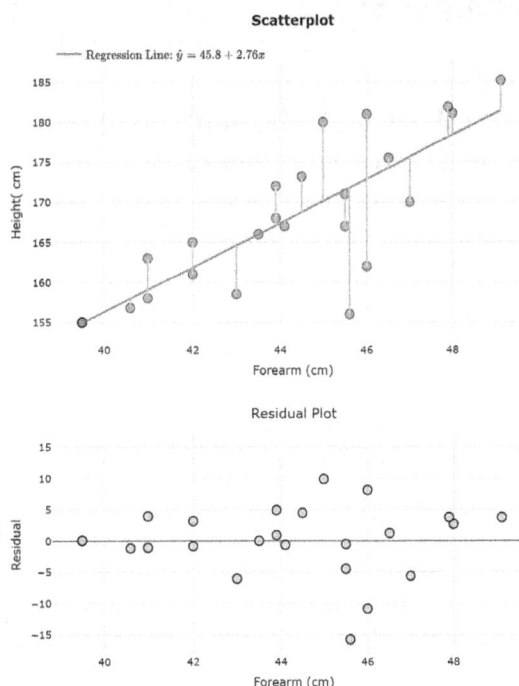

Figure 11: Least-Squares Line and Residual plot

When using a fitted model to predict a value of y (the response variable) from x (the explanatory variable), the associated prediction interval depends on the standard deviation of the residuals. This is not to be confused with a confidence interval for a mean response. The standard deviation of the residuals can be used to create a prediction interval for the height of all high school students with a forearm length of 42. Thus, for a randomly selected high school student with a forearm length of 42cm, it is predicted that they will have a height in the range from 150.1cm to 173.3cm. This prediction interval is calculated by adding and subtracting 2*(standard deviation of the residuals) to the predicted height value. This assumes that the distribution of heights for high school students with a forearm length of 42cm is bell-shaped. For these data, the standard deviation of the residuals is 5.8 (not shown here, but provided as part of the computer output), so 2*(standard deviation of the residuals) = 2*(5.8) = 11.6 cm. Adding and subtracting this amount from the predicted height of 161.7 gives a prediction interval for the height of a student with a forearm length of 42cm from 150.1 to 173.3 cm.

Interpret the Results

Students in the statistics class conclude that forearm length and height are linearly related and forearm length might be used to predict height. For example, for a specific forearm length of 42 cm, they should be reasonably confident the individual heights of the high school students will be between 149.5cm to 174.2cm. This is a fairly wide interval (just under 10 inches), which suggests that while they might use the regression equation to predict the height, they should expect quite a bit of variation and so perhaps will need to do much hemming.

Example 6: Napping and Heart Attacks – Inferring Association from an Observational Study

Formulate statistical investigative questions

The effect of napping (whether taking naps or not) on cardiovascular disease (CVD) such as heart attacks is uncertain. Some studies have indicated a positive effect of napping, some a negative, and some no effect. However, many of these studies did not consider napping frequency. A statistical investigative question to consider is:

Is there an association between napping frequency and CVD?

Collect data/Consider data

Observational studies are often the only option for situations in which it is practically impossible or unethical to randomly assign treatments to subjects. Such situations are a common occurrence in the study of causes of diseases. As researchers consider how to design a study to answer this statistical investigative question, it would be impossible to randomly assign subjects to treatment groups telling subjects they cannot take weekly naps, to take exactly 1-2 weekly naps, or to take more weekly naps and then trust subjects to honor those number of weekly naps for a long observation period.

Instead, researchers in Switzerland (Hausler, Haba-Rubio, Heinzer R, Marques-Vidal, P., 2019) designed an observational study following 3,462 Swiss subjects who had no previous history of CVD. The subjects were asked to self-report their typical nap frequency during a week.

Table 8: Nap Frequency and CVD event

CVD event	Nap Frequency				TOTAL
	No nap	1-2 weekly	3-5 weekly	6-7 weekly	
Yes	93	12	22	28	155
No	1921	655	389	342	3307
TOTAL	2014	667	411	370	3462

week. The researchers then classified the subjects into four groups OF the explanatory variable. No nap, 1-2 weekly, 3-5 weekly, 6-7 weekly. The response variable was whether or not (yes or no) the subject had a CVD event during the next 5 years. Table 8 shows the summary counts of subjects for the categories of the explanatory and response variables at the end of the 5-year period.

Analyze the Data

Understanding the nature of the counts in Table 8 is essential for analyzing the data. Each observation is accounted for in one of the interior cells that show joint frequencies (where categories for the two variables overlap). Additionally, each observation is one of the total 3462 observations as well as one of the row marginal frequencies and one of the column marginal frequencies. For example, a subject who is classified as no nap and no CVD event is one of the 1921 joint frequency number and also one of the 3307 subjects who had no CVD and one of the 2014 subjects who do not nap. Analysis of the table might lead students to examine whether there is a noticeable difference in the percentage of subjects who had a CVD event for each napping frequency category. For example, out of the subjects who took no weekly naps, the percentage who had a CVD event was 93/2014 = 4.6%. Of those subjects who took 1-2 weekly naps, the percentage who had a CVD event was 12/667 = 1.8%. Table 9 displays all these conditional percentages.

Table 9: Conditional Percentages on CVD event for each Nap Frequency

CVD event	Nap Frequency				TOTAL
	No nap	1-2 weekly	3-5 weekly	6-7 weekly	
Yes	4.6%	1.8%	5.4%	7.6%	4.6%
No	94.4%	98.2%	94.7%	92.4%	94.4%
TOTAL	100%	100%	100%	100%	100%

The different napping groups show differences in percentages for the incidence of a CVD event. The 1-2 weekly group had the lowest incidence percentage at 1.8%, which was at least 3 percentage points lower than that of the other three napping groups. The group who took no naps had an incidence percentage of 4.6%. The group of subjects who took 6 - 7 weekly naps had the highest percentage of CVD events at 7.6%. These results are shown in the percentage bar graph in Figure 12.

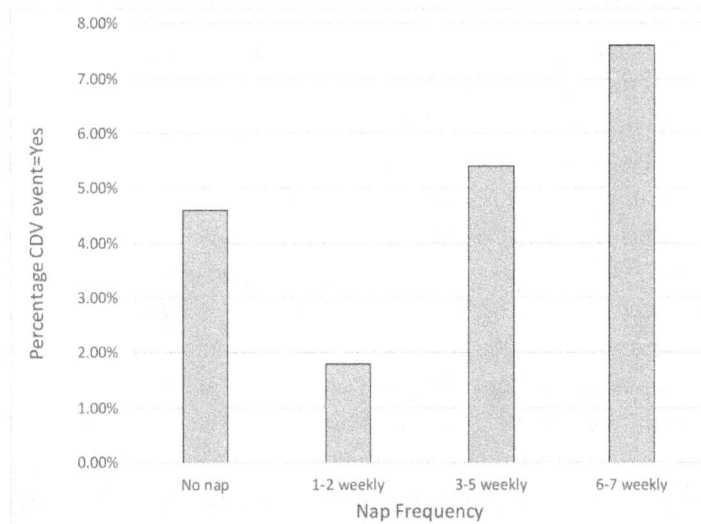

Figure 12: Conditional Percentages for CVD event = Yes

Descriptively, it appears that those who took no naps are at moderate risk of CVD within 5 years, those who take 1-2 naps are at the lowest risk, and the risk increases the more naps the person took. (We're using the conditional percentage of having a CVD given the number of naps as a measure of risk.) But could this association be due merely to chance? In other words, might there be no relationship between the number of naps and the occurrence of CVD?

One approach is a Chi-squared test, which compares the number of counts actually observed in each cell with the number of counts expected in each cell if there were no association between the two variables. Whereas understanding the technical underpinnings of

Table 10: Expected values based on conditional percentages

CVD event	Nap Frequency				TOTAL
	No nap	1-2 weekly	3-5 weekly	6-7 weekly	
Yes	90.17	29.86	18.40	16.57	155
No	1923.83	637.14	392.60	353.42	3307
TOTAL	2014	667	411	370	3462

Chi-squared tests are beyond Level C, students can use technology and interpret the results in context. The expected values can be computed by using the marginal percentages (CVD Yes = 155/3462 = 4.48%; CVD No = 3307/3462 = 95.2%) and multiplying them by the column totals resulting in the following:

The Chi-squared statistic is 20.3 with a p-value of 0.0001 and 3 degrees of freedom.

Interpret the results

The Chi-squared test resulted in a p-value of 0.0001. There is a 0.01% probability of getting a Chi-square test statistic of 20.3 or larger if what is being observed is due solely to chance under the presumption of no association. Therefore, there is

Chi-Squared Distribution with df = 3

H_0: Independence, X^2 = 20.28, df = 3, P-value = 0.0001

0.9999   0.0001

$X^2 = 20.28$

Figure 13: Chi-squared results from applet

convincing evidence to conclude there is an association between the number of naps taken and the occurrence of CVD. Note that the chi-squared approach does not provide information about the direction of the association; it doesn't tell whether more naps is associated with an increased risk or a decreased risk. In fact, from considering the bar chart, the association may be non-linear. The risk is higher for those taking no naps than for those taking 1-2 naps weekly, and the risk then increases as the number of naps increases. The risk for the no-nap group compared to the 1-2 naps weekly group could be measured by the relative risk (RR) computed as RR = 4.6%/1.8% = 2.55. We can interpret this to mean that the percentage of CVD events is 2.55 times larger for no nappers than for 1-2 weekly nappers.

Does this study imply everyone should start taking 1-2 naps a week to avoid CVD events and not take more than 2 naps a week? Not necessarily, because this was an observational study, a cause-and-effect statement cannot be made. There may have been some confounding variables other than nap frequency impacting the response variable of CVD event. For example, if older people are more likely to take 6-7 naps per week and are also more likely to have CVD, then the variable of age might be important to consider. In this instance, there would be an association between napping frequency and CVD even if napping has no effect on CVD. Likewise, if people with other health issues are more likely to nap and more likely to have CVD, then having other health issues is confounded with napping frequency.

When experiments cannot be conducted and observational studies are used instead, it is important to identify potential confounding variables. Then, if possible, control for these variables. One way to do this is to disaggregate the data to analyze the explanatory and response variables by the categories of the confounding variable. As students advance beyond Level C, they will learn more about how to control for potential confounding variables with the design of the study. They will also learn more formal statistical methods for analyzing the data, such as confidence intervals for differences, and how to judge the practical importance of the association by evaluating effect sizes.

Example 7: Working-age Population – Working with Secondary Data

Formulate statistical investigative questions

Students and the general public routinely encounter statistical graphics that are rarely seen in Pre-K–12 classrooms. Part of statistical literacy is being able to grapple with data presented in the media using various unconventional representations.

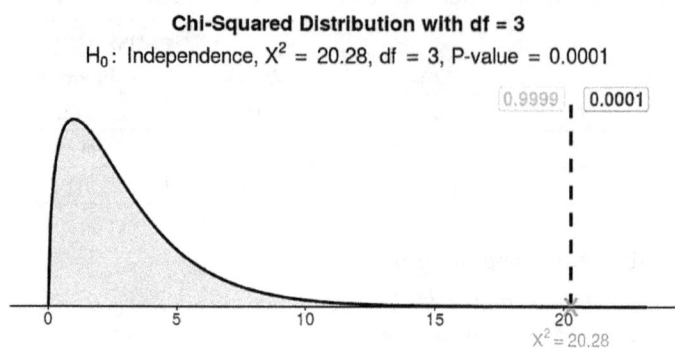

The *New York Times* has a weekly feature, What's Going On in This Graph? (WGOITG) (https://www.nytimes.com/column/whats-going-on-in-this-graph) that provides excellent examples of such graphics. Each week, a statistical graphic that appeared in an earlier story is presented with some context removed. Students are invited to post responses to three prompts: *What do you notice? What do you wonder? What's going on in this graph?* Alternatively, students might be asked to write a catchy headline that captures the graph's main idea.

For example, one WGOITG was on the topic of changes in the size of the working-age population over a 10-year period. In a context such as this, posing an answerable statistical investigative question is the primary focus. After examining this particular graph, students might pose the following statistical investigative question:

How has the working-age population shifted across the United States?

Collect data/Consider data

The graphic given in WGOITG is a heat map, which represents a statistical analysis of the data. The legends provided by the analyst help us understand the data themselves. The graph color-codes each county in the United States according to the shift in the working-age population. The graph was converted to grayscale for this publication. For example, a county is colored dark grey if the population of working-age people (defined as 25 to 54 years old) has decreased by more than 10% from 2007 to 2017. On the other end of the color spectrum, a light grey color represents a county that has increased in the working-age population by over 10% from 2007 to 2017.

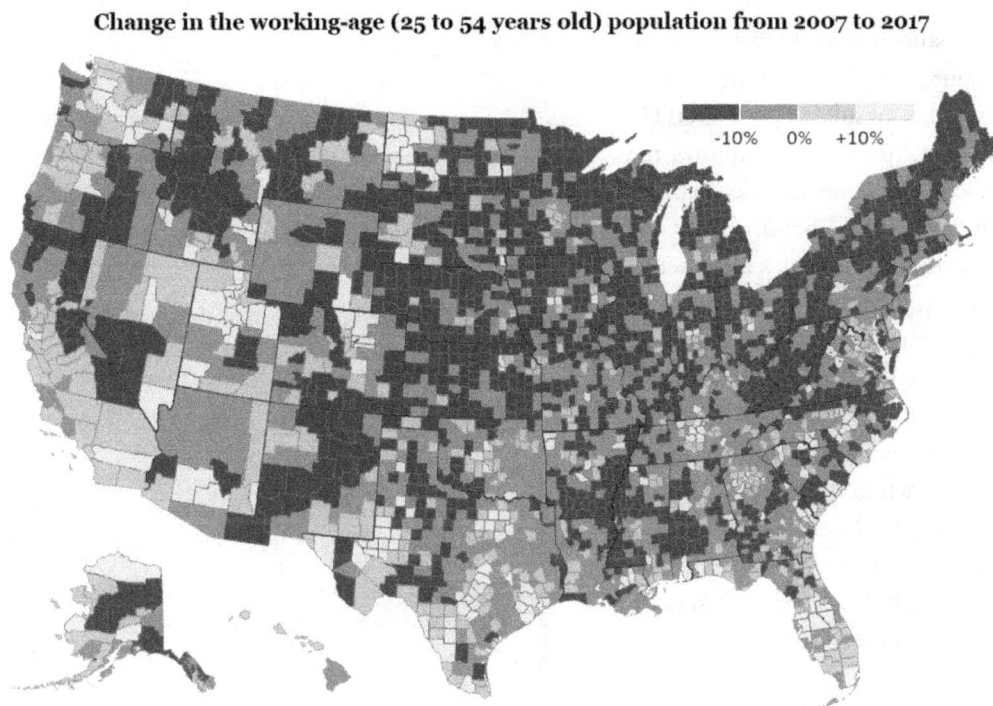

Figure 14: Heatmap of the United States showing change in population.
Source: *The New York Times's* What's Going On in This Graph? September 19, 2019
© 2020 The New York Times Company

Students in Level C need to study summarized secondary data provided in graphics in order to understand the multiple variables being displayed. Level C students should be able to identify the observational units (counties) and all variables represented in this graph. For example, one variable represented is the percent change in the number of people aged 25-54 years who reside in a county. The data set that produced these graphics might have had columns labeled "county", "working_age_population_2007", "working_age_population_2017", "percent_change".

Analyze the data

In this case, the analysis has been done. The graphic is the analysis of the data. This is typical when interpreting others' work as it appears in news media or scientific publications.

Interpret the Results

Level C students may notice that the middle and north-eastern parts of the country are predominantly dark grey, suggesting a decrease in the working-age population in those parts of the country, while the west and south have tended to see an increase in the working-age population. Overall, the data suggest large shifts over time in the number of working-age individuals. Students might also comment on the within-state variability. Some states, such as Nebraska, Maine, or Vermont, are almost entirely dark grey, indicating that the entire state suffers from a decrease in the working-age population. Other states, such as Utah, are mostly light grey, indicating large increases in the working-age population. Other states, such as Texas and California, have more variability; some areas of the state see increases while others see declines.

Level C students might also consult other sources of information to help interpret their results. The WGOITG heat map pictured above captures the difference over one decade without considering the month-by-month or year-by-year time increments. Considering the monthly data provides an opportunity to further interrogate the data. The time series graph (Figure 15) shows the number of working-age individuals in the entire United States for each month from January 2007 through December 2017. The working-age population declined overall through 2014 and then appears to have been increasing since 2015.

A Level C student might notice the pattern in the timeplot where there is a drop from the end of one year to the beginning of the next (e.g., the count in Dec 2007 is about 126 million but

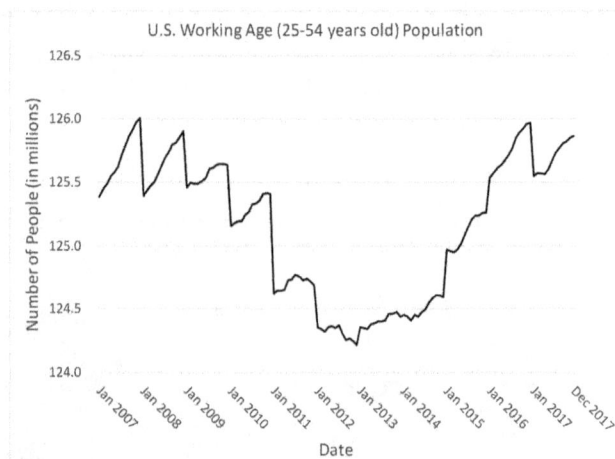

Figure 15: Timeplot of the monthly data from the heatmap

Figure 16: Timeplot with a moving average added

falls in January 2008 to less than 125.5 million). Students might speculate that the age of a person was calculated based on their birth year rather than birth month. In order to account for this but not lose the level of detail available in the monthly data, they can compute a moving average. Moving averages can vary in the time period used (e.g., 3-months, 6-months, etc.); in this case, it makes sense to use a 12-month moving average, which will smooth out these annual adjustments. The graph in Figure 16 includes the monthly data as well as the 12-month moving average.

If monthly data from 2007 to 2017 are available, the first 12-month moving average that can be calculated is for June 2007 (includes Jan 2007 to Dec 2007) and the last value that can be calculated is for June 2017 (includes Jan 2017 to Dec 2017). The moving average has a similar but smoother trend. There are other ways to smooth patterns of graphs, such as seasonality adjustments and exponential smoothing.

Example 8: Classifying Lizards – Predicting a Categorical Variable

Formulate Statistical Investigative Questions

Level C students go beyond Level B by tackling more open-ended questions and by considering datasets that might require some preparation before analysis. Data sets appropriate for Level C may have additional variables that are not used in the analysis, forcing students to consider different aspects of the data and how multiple pathways might be productively pursued.

Suppose that students in a science class are exploring the impact of human development on wildlife. In an earlier analysis, students discovered that lizards in "disturbed" habitats (habitats with substantial human development) tended to have greater mass than lizards in natural habitats. Students posed this statistical investigative question:

Can a lizard's mass be used to predict whether it came from a disturbed or a natural habitat?

Students can approach this statistical investigative question by trying to define rules for how a randomly selected lizard can be classified as coming from either a disturbed or natural habitat.

Collect data/Consider data

A biologist captured a number of individuals from one species of lizard, Anolis sagrei, across two different habitat types on each of four islands in the Bahamas. The data was collected by Erin Marnocha during her time as a graduate student at the University of California Los Angeles. The lizards were selected through a random sampling procedure. Some of the lizards lived in habitats that were developed by humans (disturbed) while others lived in natural habitats with no human development (natural).

A total of 160 lizards were captured, 81 from natural habitats and 79 from disturbed habitats. Once captured, the lizards were measured across several physical characteristics, including their mass, length, breadth, foot span, head width, etc. Lizards captured from habitats with human development were labeled "disturbed" and those captured from habitats without development were labeled "natural."

The first few rows of the data are given in Figure 17. Note that the data are displayed in long format where each row represents a lizard, and the columns tell us the measured characteristics of the lizard.

To consider and understand these secondary data, Level C students can examine and query the dataset using technology. Level C students should note that although there are 160 lizards included in the data set, quite a few of them have missing values for one variable or another. The "Tail length" variable is missing quite a few values (for example, see Index 3 in Figure 17). Students with experience catching lizards will know why: lizards can lose their tails to help evade capture.

index	Island	Habitat	Mass (g)	SVL (mm)	Hindlim... ngth (mm	Hindspan (mm)	Forelimb (mm)	Forespan (mm)	Gape width (mm)	Head depth (mm)	Toe pad width (mm)	Tail length (mm)
1	New Pr...	Natural	2	49	11.2	23.3	7.1	20.5	7	4	0.9	
2	New Pr...	Natural	2.2	45	10.4	25.4	7	20.8	7.6	5.1	1.2	78
3	New Pr...	Natural	2.4	50	11.6	27.1	7.1	22.5	7.2	4.6	1	
4	New Pr...	Natural	2.6	48	11.2	25.8	7.2	21.4	7.8	4.5	1.1	83
5	Exuma	Natural	2.7	53	11.4	27	7.3	22.9	7	5.1	0.8	
6	New Pr...	Natural	2.7	48	11.3	25.1	8.1	21.9	7.2	4.5	1.2	
7	Harbor I...	Natural	2.7	47	10.8	24.2	7	19.9	8.5	4.7	0.9	83
8	Eleuthe...	Natural	2.7	46	10.9	25	7.3	21	7.5	4.6	1.1	
9	Exuma	Natural	2.8	49	12.2	25.8	7.2	22.6	7.4	4.4	1.1	94
10	New Pr...	Natural	2.8	49	11.9	25.7	8.7	23.2	8	4.5	0.9	89
11	New Pr...	Natural	2.8	51	11.4	25.3	7.6	22.7	8.1	4.8	1.1	
12	New Pr...	Natural	2.8	50	12.1	26.4	7.6	22.5	7.4	4.4	1.2	
13	New Pr...	Natural	2.9	51	11.8	27	8.8	23.3	8.7	5.2	1.1	90
14	Eleuthe...	Natural	2.9	48	11.5	25.8	7	21.6	8.1	4.6	1	
15	New Pr...	Natural	3	51	10.4	26.2	8.1	22.5	7.3	4.3	1.1	86
16	New Pr...	Disturb...	3	52	11.6	25	7.5	21	7.7	4.1	0.9	

Figure 17: Beginning of lizard data in long format

This statistical investigative question relies on only two variables: "Habitat" and "Mass" (measured in grams). For ease of analysis, students might create a new data set that contains only those two variables.

Analyze the Data

A comparative dotplot of the mass (Figure 18) shows considerable overlap between the two types of habitats. The statistical investigative question focuses on predicting whether a randomly selected lizard comes from a natural or disturbed habitat (categorical response variable) using the mass of a lizard as the explanatory variable. In Level C, students can approach the analyses using classification (beyond Level C, several additional modes of analyses exist to help answer such statistical investigative questions).

A classification approach essentially requires students to propose a rule based on the mass of the lizard that predicts the type of environment the lizard came from. For example, consider a potential rule that classifies lizards <6.25 g to be from natural habitats ("natural"). Students in Level C can analyze whether this provides a good prediction rule.

The dotplot shows that if lizards weighing less than 6.25g are classified as from natural habitats then 100% of these lizards are correctly classified (because all of these lizards in our sample weigh less than 6.25g). However, this is not a perfect classification rule, because a high percentage ($53/79 = 0.671 = 67.1\%$) of the lizards that are from disturbed habitats would be misclassified because they weigh less than 6.25 g. Since the dotplots show a lot of overlap in the masses of the two groups, the classification of a lizard as coming from a disturbed or a natural habitat will not be straightforward.

Level C students should recognize that less overlap between distributions provides a greater classification success rate. Also, changing the cut-off point can vary the misclassification rate. For example, if lizards with mass less than 5.0 g are classified as "natural," fewer errors will be made when encountering lizards that are truly from disturbed habitats, but more errors will be made with lizards from truly natural habitats.

Analyzing the data using a classification approach is obtainable for Level C students. With appropriate technology or careful counting of dots in the dotplot, students can compute and interpret a confusion matrix for any cut-off rule they choose. A confusion matrix shows the true values in the columns and the classification categories in the rows (or vice versa). Thus, the upper-left cell of the matrix below indicates that 26 lizards were correctly classified as belonging to a disturbed habitat. A perfect classifier will have 0's on the off-diagonal.

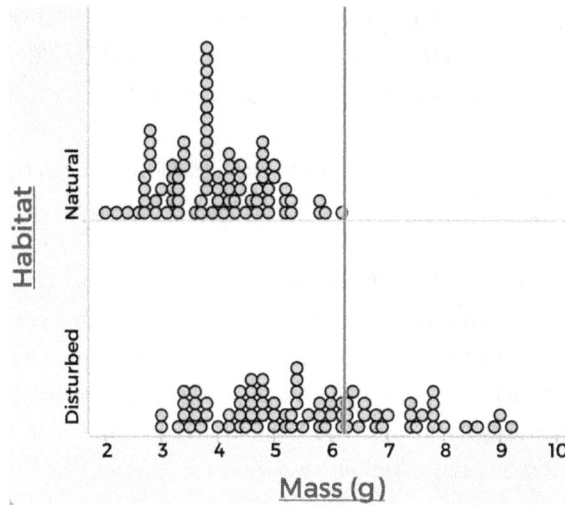

Figure 18: Stacked dotplots showing mass of 6.25 g as cutoff

The first confusion matrix in Table 11 shows the results of classifying with the rule where lizards with mass less than 6.25 g are classified as "natural." All 81 of the lizards from natural habitats were classified as natural, but 53/79 lizards from disturbed habitats were incorrectly classified. Overall, the misclassification rate is (number of lizards misclassified)/ total number of lizards = (53 + 0)/(26+53+0+81) = 0.331 or about 33%.

Table 11: Confusion Matrix for mass < 6.25 g

	Truly from Disturbed Habitat	Truly from Natural Habitat
Classified as "disturbed"	26	0
Classified as "natural"	53	81

By changing the rule to classify a lizard as natural if its mass is less than 5 g, a new confusion matrix results:

Table 12: Confusion Matrix for mass < 5 g

	Truly from Disturbed Habitat	Truly from Natural Habitat
Classified as "disturbed"	49	11
Classified as "natural"	30	70

Now the number of misclassified lizards is 30+11 = 41. Therefore the misclassification error rate is (30 + 11)/(49+30+11+70), or about 26%. The new rule improves on the previous rule.

Students can continue to explore classification rules to determine which produces the lowest misclassification rate.

Determining an optimal procedure for classification is beyond Level C. Instead, students in Level C should understand how changes in a rule affect misclassification rates by computing probabilities. Level C students should understand that a classification approach can be used to answer statistical investigative questions that focus on predicting a response variable that is categorical by using another variable. Beyond Level C, several other approaches exist to help answer such questions (e.g., logistic regression, random forests, machine learning, and deep learning algorithms). In general, classification is part of prediction—here, for example, one is trying to predict the category of a randomly selected lizard based on one or more variables. These methods are often discussed in the news, especially with the advances in machine/deep learning. For example, facial/image recognition is an advanced example of a classification problem prevalent in the media today (e.g., https://www.nytimes.com/2019/12/19/technology/facial-recognition-bias.html). To promote statistical literacy, students should learn the basic ideas of classification approaches in Level C.

At Level C, given data about a new lizard from the population, students should be able to classify the lizard and provide the probability of misclassification for that new observation.

Interpret the Results

A misclassification rate of 26% might strike some students as good and others as not-so-good. Students might wonder how to evaluate misclassification rates. To find a "baseline" rate for comparison, one approach is to ignore the mass of a lizard and classify all lizards into the largest group. Because there are 79 lizards from disturbed habitats and 81 from natural habitats, all lizards can be classified as "natural" habitats without looking at their mass.

This rule will misclassify all of the lizards from disturbed habitats, so the misclassification rate will be 79/(81+79) = 0.49 or 49%. By comparison, the previous classification rule with a 26% misclassification error rate (lizards with a mass of less than 5.0 g are "natural") is quite a bit better than this baseline rate.

Because these lizards were randomly sampled, these misclassification rates can be interpreted as estimates of the proportion of incorrect classifications for future captured lizards, as long as they are captured from the same population as the lizards in this dataset. Level C students can conclude that, in fact, lizards can be classified based on their mass in such a way that the probability of misclassification is lower than if they had been simply classified as all belonging to the same habitat.

Analysis Revisited

The previous approach depends on a single explanatory variable – mass of the lizard. How might multiple variables be used to classify the lizards? More specifically, the statistical investigative question can be rephrased as:

Can a lizard's mass, head depth, and hind limb length be used to predict whether it came from a disturbed or a natural habitat?

One approach that furthers the simple classification approach which is accessible to Level C students if they have the appropriate technology is to apply a Classification and Regression Tree (or CART). CART is an example of a modern statistical approach that relies on an algorithm, rather than a mathematical model. An "algorithm" in this case is taken to mean a series of rules.

Although applying and interpreting the CART algorithm is accessible to Level C students, the details of the algorithm are not necessary for students to begin their understanding. CART begins with splitting the data according to a rule. Suppose the rule is the one above: classify a lizard as natural if its mass is less than 5g. This rule splits the data into two groups.

One group consists of lizards less than 5 g that are classified as "natural" (even though this group may contain some that are not from the natural habitat), and the other group consisting of lizards larger than 5 g that are classified as "disturbed" (even though some may not be from the disturbed habitat).

Students can then consider the group of lizards classified as "natural" and determine if, based on one of the variables in the data set, they can further split this group into two smaller groups.

For example, the initial classification split classified 100 lizards as "natural" because their mass was less than 5 g (this can be seen in the confusion matrix in Table 12 as 30+70=100). Now, considering only these 100 lizards, students can repeat the process with a new variable, for example by making stacked dotplots of the 100 lizards' head depth measurement. The lizards truly from a disturbed habitat would be plotted on top and the natural lizards below. The result is shown in Figure 19.

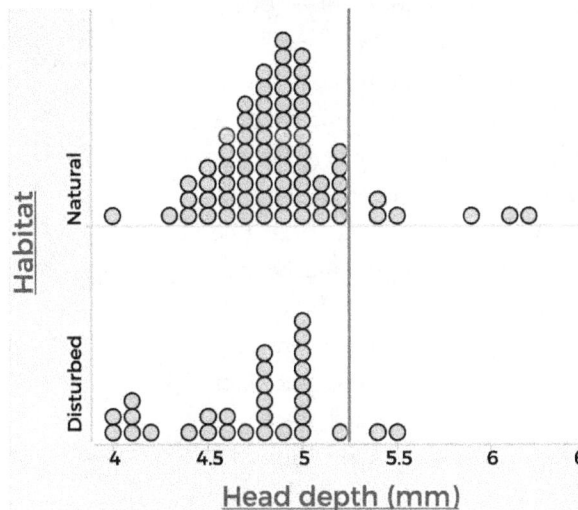

Figure 19: Stacked boxplots for 100 lizards with mass less than 5 g showing 5.25 cm head depth as a cutoff

Again, students can consider different "cutoff" values. A first attempt might be to draw a cutoff at a head depth of 5.25 mm, since visually that seems to separate the bulk of the data on the left from the remaining data on the right. Because the natural lizards have a slightly larger mean head depth, students may classify lizards to the right of the cutoff line shown in Figure 19 as "natural," and those to the left as "disturbed." This results in a new set of classification rules:

1) If the mass is less than 5 g, consider the head depth. If the head depth is less than 5.25 mm, classify as "disturbed." If the head depth is greater than or equal to 5.25 mm, classify as "natural."

2) If the mass is greater than or equal to 5 g, classify as "disturbed."

The next step would be to repeat this procedure on the 60 lizards whose mass was greater than or equal to 5 g and who were thus classified as "disturbed."

This process can continue for quite a while. At each step, any variable may be considered in order to split the data into two classification groups. The CART algorithm does what no human would be patient enough to do. At each step of the process, it considers all available variables and all possible cutoff values. It determines which provides the lowest misclassification rate, and then it splits the data based on that rule. This creates two new groups, and the process is repeated on those two groups. This, of course, produces four groups, and the process is repeated on each of those four new groups. The process stops when performing a split provides no improvement in the misclassification rate or when a group is too small to split.

The result could be written as a series of rules, but it is better visualized as a tree. Figure 20 shows the CART for these data. Each split contains a rule stated as a condition. If the condition is true, move down the left branch and continue in this fashion until reaching an end node. The node indicates how to classify the observation.

For example, consider a lizard with a mass of 4.5 g, head depth of 5 mm, and a hind limb measurement of 11 mm. The first rule (mass ≥ 5.35 g) does not apply to this lizard, so this tells a student to move to the right branch. The next rule considers head depth, and because the lizard's head depth is not less than 4.25 mm, a student can again move to the right. Next, the mass is considered again, and since it is greater than 4.25 g, this time the move is to the left. Finally, because the hind limb measurement is less than 12.65 mm, the move is to the left, which classifies this lizard as "disturbed." Had the hind limb length been greater than 12.65 mm this lizard would have been classified as "natural."

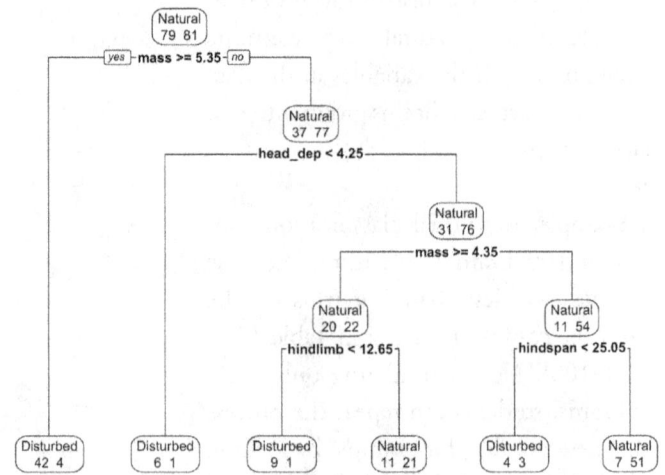

Figure 20: CART Tree for Lizard data where the left branch is taken when rule applies.

The numbers below the classification label indicate the number of lizards of both types that were classified with that category label. For example, in the end node shared by our hypothetical lizard (Disturbed 9/1), nine of the lizards from our sample data who were sent to that node were truly from a disturbed habitat, and one was truly from a natural habitat. So, for lizards with traits similar to these, the misclassification rate is 1/10, or about 10%.

Interpret the Results

Each split of the tree is a decision, and each lizard will be classified according to the end-nodes of the tree. Of the two numbers in each node, the number on the left represents the number of lizards truly from disturbed habitats, and the number to the right is truly from natural habitats. For example, the left-most node classifies lizards as disturbed. This classification is correct for 42 lizards and incorrect for 4 lizards.

The overall misclassification rate for this tree is 18% (27 out of 160 lizards), a figure computed by the software that produced the tree. We could confirm this by counting the number of misclassifications and dividing by 160, the total number of lizards. For example, the left-most node is labeled as "Disturbed" and contains 42 lizards from disturbed habitats and 4 from natural habitats, and thus there are 4 misclassifications. Continuing left to right, theare are an additional 1, 1, 11, 3 and 7 misclassifications for a total of 27 misclassifications out of 160 lizards. This gives the misclassification rate of 27/160 or about 18%. This tree improves upon the earlier classification rule, which was based solely on mass.

If technology is not available to create trees, students can still benefit from creating their own trees using sets of rules of their own design.

Summary of Level C

Students at Level C should become adept at using questioning throughout the statistical problem-solving process. In Level C, statistical investigative questions expand from summative and comparative situations to include questions about associations and relationships among multiple variables, including predictions.

Once an appropriate plan for collecting data to answer a statistical investigative question has been implemented and the resulting data are in hand, the next step usually is to summarize the data using tools such as graphical displays and numerical summaries.

At Level C, students should be able to select data analysis techniques appropriate for the type of data available, produce descriptive statistical analyses, and describe in context the important characteristics of the data. Level C students should be able to provide more sophisticated and insightful interpretations compared to Levels A and B. These interpretations should integrate the context and objectives of the investigation to draw conclusions from data and to support these conclusions using statistical evidence (both descriptive and inferential).

Level C students should see statistics as providing powerful tools that enable them to answer statistical investigative questions and to make informed and trustworthy decisions. Students also should understand the limitations of conclusions based on the data from sample surveys and experiments, and they should be able to quantify the uncertainty associated with these conclusions using margin of error and related properties of sampling distributions. Statistically proficient students at Level C are "healthy skeptics" of statistical information, capable of asking questions to assess the validity of statistical findings. At the same time, they are cognizant of the power of statistical analyses to find information and meaning in data. Level C students should understand that because their own work will be read by others, they should carefully document both their data and their analyses so that others can reproduce their analysis. Level C students should understand the ethical consequences of their experiments and analyses.

While Level C rounds out students' statistical literacy, students should understand that more complex statistical investigative questions, data collection methods, study designs, and analysis methods exist. Beyond Level C there is an additional wealth of statistics to be learned. Levels A, B, and C material provides students with the basics of statistical literacy at this point in time. The material will continue to evolve, however. It is not just data that is changing, it is processes, technology, philosophies, strategies, and many other changing facets that lead to the evolution of statistical methods. Statistics is not a stagnant discipline; instead, it is an evolving field that is rooted in the statistical problem-solving process discussed in this report.

Assessment

Assessments provide feedback to learners, teachers, and guardians by providing a gauge of what students know and can do at a particular point in time. It is essential that instruction and assessment be aligned, consistent with the principles outlined in the different Levels in GAISE II. Assessments of students' statistical thinking should assess conceptual understanding, be set in a context, and require interpretation (Peck, Gould & Miller, 2013). These criteria apply to formative and summative assessments regardless of the form of assessment (e.g., discussions, activities, projects, reflective essays, homework problems, tests). While it may not be feasible to assess statistical reasoning at all stages of the statistical problem-solving process in a single item, it is important to construct items that assess multiple parts of the process across time.

Many statistics items commonly found on standardized or locally developed assessments require lower-level cognitive skills (such as recall) and focus on procedures and definitions. In addition, many of these items do not focus on statistical reasoning but rather focus on mathematical computations. For example, a task asking students to find the mean for a set of numbers is a low-level task that involves mathematical computation and no statistical reasoning. Even a task that involves giving students all but one data point and the mean and asking them to find the missing data point is a lower-level mathematics task. Fortunately, there are several examples of robust assessment items on standardized assessments as well as resources for teachers to find such items for classroom use.

National and International Standardized Assessments

Several national and international standardized assessments include a focus on data analysis and statistics. For instance, the Programme for International Student Assessment (PISA), which is given to 15-year-old students around the world to assess ability to apply knowledge to real-world situations, is expected to focus on mathematics in 2021 with computer simulations and conditional decision making identified as topics for special emphasis. According to the College Board, 29% of the items on the mathematics portion of the SAT address problem solving and data analysis (which includes proportional reasoning and probability). Data analysis, statistics, and probability is one of five content strands assessed on the National Assessment of Educational Progress (NAEP). And, of course the Advanced Placement (AP) Statistics Examination contains only questions about statistics. The standard exam format includes 40 multiple-choice questions, five short free-response problems, and 1 investigative task, and over the years it has progressed to have an emphasis on conceptual understanding and interpretation rather than simply on arithmetic calculation. Selected items are released after each administration and are available to teachers.

Sources of Quality Items for Educators

Creating high quality assessment items is time-consuming and challenging. Thus, we profile two sources that have been vetted and can serve as resources for teachers – Levels of Conceptual Understanding in Statistics (LOCUS) and Statistics Education Web (STEW).

LOCUS
The LOCUS assessments are based on the original GAISE I framework, aligned with the Common Core State Standards, and constructed as reliable measures of understanding across the levels of development

and throughout the statistical problem-solving process. Both multiple choice and constructed response questions are included, and there are different versions of the assessments based on the different GAISE Levels. Each version has equated forms for pre/posttest purposes, so they can be used for both formative and summative assessments.

Both types of questions, multiple choice and constructed response, can be viewed by grade level or by component of the statistical problem-solving process. Multiple choice sample questions include associated standards, student performance, correct answers, and commentaries. Constructed response questions include the additional information of scoring rubrics, common misunderstandings, sample responses, and resources.

STEW

STEW online resource includes peer-reviewed lesson plans. STEW lesson plans follow a standard format and include objectives, applicable Common Core State Standards, and instructions for enacting the lesson in the classroom. Although they are sorted by grade level, many lessons can be modified to address similar content at different levels. Material may also include problems with sample solutions, slides that can be used to guide class discussions, and directions for technology tools. Educators are encouraged to use the lesson plans that are on this site as well as contribute new lesson plans that they have found effective in their classrooms. These lessons could be adapted to assessments.

Some of the following examples of assessment items are from the resources mentioned (e.g., LOCUS and STEW) and they are provided by Level in the following sections. These examples are meant not to be exhaustive but instead showcase how statistical thinking can be assessed.

Level A Assessment Examples

Example 1: Measures of center

The students in Ms. Kieffer's class kept track of how many children in their class ate school lunch during the month of February. Table 1 shows what they found:

Table 1: Number of children eating a school lunch

M	T	W	Th	F
18	21	19	20	22
25	17	19	18	19
19	20	21	21	23
21	20	23	19	23

If the principal asked you approximately how many people in your class ate the school lunch each day, what would you tell her? Explain how you obtained your answer.

Example 2: Variability, sampling, and drawing inferences

I want to know what the favorite food is of people in the world. Would I get a reasonable answer if I used your class as a sample? Why or why not?

Example 3: How to model data in a variety of ways to see how the conclusions drawn are influenced by the way the data are represented
(**from "Candy Judging" Lesson Plan on** STEW website)

Jeremy's class rated the chocolate candies from most favorite (1) to least favorite (4). His class data are recorded and shown in Table 2.

Table 2: Ranking from most favorite (1) to least favorite (4)

	Special Dark®	krackel®	mr. Goodbar®	Milk Chocolate
Jeremy	1	2	3	4
Kayla	4	2	3	1
Quentin	1	2	3	4
Ken	4	3	1	2
Jake	1	3	4	2
Polly Ann	1	2	3	4
Rocco	1	3	2	4
Drake	4	2	1	3
Corrine	1	2	3	4
Kris	4	2	1	3
Mary	4	3	2	1
Casey	1	3	4	2
Mel	4	3	2	1
Lisa	1	3	2	4
Cindy	4	2	1	3

Use the approaches below to determine the class favorite.

(a) Find the chocolate that was selected as favorite (ranked with a 1) the most often.

> *Represent this data in a picture graph (or dotplot). Which candy would represent the class favorite if this method were used to determine the favorite?*

(b) Find the chocolate that was selected as least favorite (ranked with a 4) the most often.

> *Represent this data in a picture graph (or dotplot). Which candy would represent the least favorite of the class if this method were used to determine the least favorite?*

(c) Find the sum of the scores of the rankings. Which chocolate would represent the favorite if the sum of the class scores was used to determine the favorite? Which chocolate would be the least favorite using this method?

(d) Find the median score of each type of chocolate. Which chocolate would represent the favorite if the median was used to determine the favorite? Which chocolate would represent the least favorite if the median was used to determine the favorite?

(e) Draw a bar graph of the score distribution for each candy. Which chocolate do you think would represent the favorite if you use the bar graphs to compare them?

Example 4: Variability
(from "Describing Distributions" Task on Illustrative Mathematics website)

Data Set 3 consists of data on the number of text messages sent in one month for 100 teenage girls who have a cell phone. Data Set 4 consists of data on the number of text messages sent in one month for 100 teenage boys who have a cell phone. Histograms of the two data sets are shown in Figure 1 (upper level A students can answer these questions).

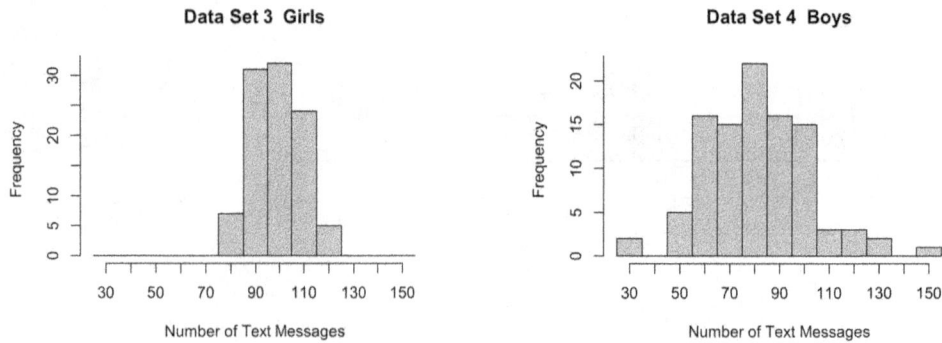

Figure 1: Histograms for Girls and Boys

(a). *Describe the data distribution of number of text messages for the girls (Data Set 3). Be sure to comment on center, spread and overall shape.*

(b). *Are Data Set 3 and Data Set 4 centered in about the same place? If not, which one has the greater center?*

(c). *Which of Data Set 3 and Data Set 4 has greater spread?*

(d). *On average, did the girls (Data Set 3) or the boys (Data Set 4) send more text messages?*

Level B Assessment Examples

Example 1: Randomness
(from LOCUS Project)

Students wanted to investigate whether the distance a male student can jump is affected by having a target to jump toward. The students decide to perform an experiment comparing two groups. One group will have male students jumping toward a fixed target, and the other group will have male students jumping without a fixed target. There are 28 male students available for the experiment.

In a few sentences, describe how you would randomly divide the 28 male students to form two groups.

The task can then be extended by providing students with box plots showing the results of the experiment (see Figure 2).

Students can be asked to

Write a concluding statement to address whether the distances the male students jumped were affected by having a target. Justify your conclusion.

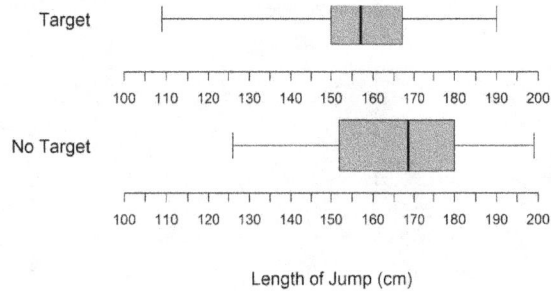

Figure 2: Boxplots for boys jumps with and without a target

Example 2: Appropriate data representations and testing conjectures

Table 3 shows data that were collected from a group of fourth graders.

Table 3: Data collected from fourth graders

Name	Age	Number of pets	Height (inches)	Weight (pounds)
Mario	9	2	50	68
Lisa	10	0	54	77
Kiko	9	1	52	73
Juan	11	0	57	83
Josef	10	4	52	71
Beatriz	10	2	55	78
Carlos	9	3	51	71
Jeannie	11	1	58	85
David	9	0	52	70
Lynn	10	3	53	75

Make a conjecture about two characteristics of children that might be related. Make a graph or chart to test your conjecture. Explain whether or not the data support your conjecture.

Example 3: Comparing two continuous quantitative data sets and draw conclusions
(from LOCUS Project)

The city of Gainesville hosted two races last year on New Year's Day. Individual runners chose to run either a 5K (3.1 miles) or a half-marathon (13.1 miles). One hundred thirty-four people ran in the 5K, and 224 people ran the half-marathon. The mile time, which is the average amount of time it takes a runner to run a mile, was calculated for each runner by dividing the time it took the runner to finish the race by the length of the race. The histograms in Figure 3 show the distributions of mile times (in minutes per mile) for the runners in the two races.

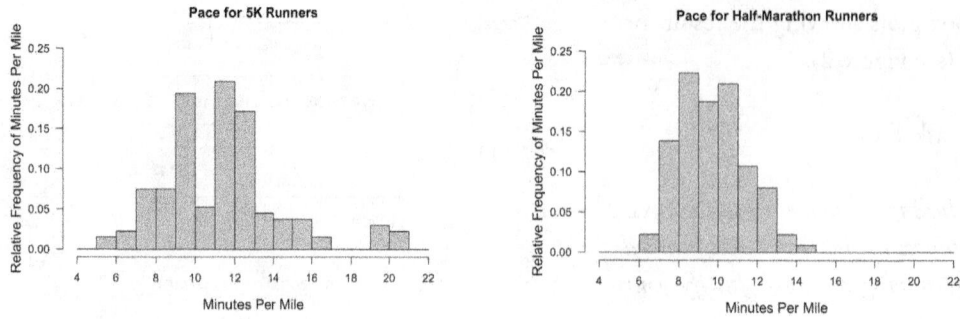

Figure 3: Histograms for runners

(a) *Jaron predicted that the mile times of runners in the 5K race would be more consistent than the mile times of runners in the half-marathon. Do these data support Jaron's statement? Explain why or why not.*

(b) *Sierra predicted that, on average, the mile time for runners of the half-marathon would be greater than the mile time for runners of the 5K race. Do these data support Sierra's statement? Explain why or why not.*

(c) *Recall that individual runners chose to run only one of the two races. Based on these data, is it reasonable to conclude that the mile time of a person would be less when that person runs a half-marathon than when he or she runs a 5K? Explain why or why not.*

Example 4: Going through the statistical problem-solving process
(from LOCUS Project **)**

The student council members at a large middle school have been asked to recommend an activity to be added to physical education classes next year. They decide to survey 100 students and ask them to choose their favorite among the following activities: kickball, tennis, yoga, or dance.

(a) *What question should be asked on the survey? Write the question as it would appear on the survey.*

(b) *Describe the process you would use to select a sample of 100 students to answer your question.*

(c) *Create a table or graph summarizing possible responses from the survey. The table or graph should be reasonable for this situation.*

(d) *What activity should the student council recommend be added to physical education classes next year? Justify your choice based on your answer to part (c).*

Example 5: Mean as a balance point and MAD
(from LOCUS Project**)**

Two soccer teams will be meeting in the city championship game. Each team played 10 games and averaged 3 goals scored per game for the season. The two dotplots in Figure 4 show the number of goals scored by each team per game for the season.

Team A   **Team B**

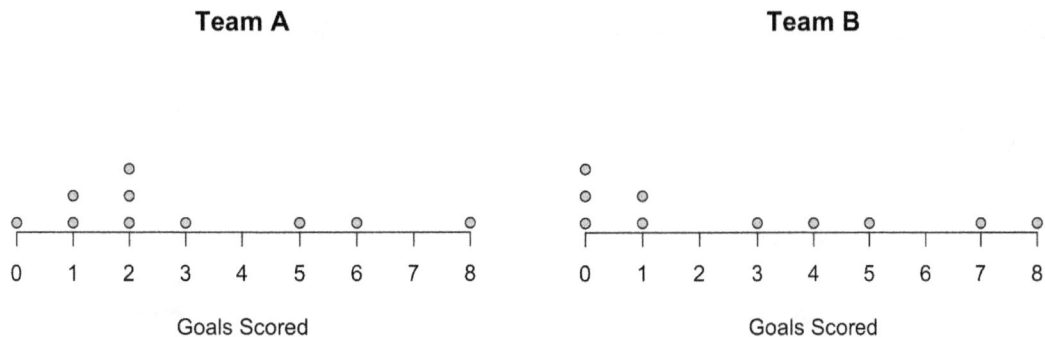

Figure 4: Dotplots for soccer goals scored

(a) *Sarah found that the mean for Team A was 3 goals by adding all the goals scored and dividing by 10. Using the data displayed in the dotplot show why 3 goals is the balance point for the goals scored by Team A.*

(b) *The MAD (Mean Absolute Deviation) for Team A is 2 goals. What does the MAD tell us about the variability in the goals scored for Team A?*

(c) *Based on the dotplots, which team has shown more variability in the number of goals scored per game over the course of the season? Explain.*

Level C Assessment Examples

Example 1: Interpreting an interval and sample size effect on the margin of error
(from LOCUS Project)

Lindsey wants to use a confidence interval to estimate the difference in the proportion of females and males at her high school who have taken an honors class. She randomly selects 50 females and 50 males from her school and asks each one if he or she has taken an honors course. Of the 50 females, 23 responded yes. Of the 50 males, 19 responded yes.

A 95 percent confidence interval for the difference in the proportion of females and males at her school who have taken an honors class is 0.08 ± 0.19.

(a) *Interpret the confidence interval in the context of this study.*

(b) *The principal at Lindsey's school is interested in the results of her study but suggests that she increase the sample sizes to 100 females and 100 males. What effect will increasing the sample sizes have on Lindsey's confidence interval?*

Example 2: Drawing conclusions about the relationship between two categorical variables
(**from** LOCUS Project)

The local health department wants to investigate whether there is an association between eating at fast-food restaurants and gender. They conduct a survey of 100 randomly selected people and ask each person the following question: "Do you eat at a fast-food restaurant at least once a week?"

(a) *What type of data (categorical or numerical) will result from the question?*

Sixty of the people who responded were men. Fifty-four percent of the 100 people surveyed eat at a fast-food restaurant at least once a week.

(b) *If there were no association between gender and eating at fast-food restaurants, what percentage of males would be expected to eat at a fast-food restaurant at least once a week? Explain.*

The survey results are displayed in Table 4. Respondents are classified by gender (male or female) and whether or not they eat at a fast-food restaurant at least once a week.

Table 4: Responses to survey by gender

Eat at a Fast-Food Restaurant at Least Once a Week			
Gender	Yes	No	Total
Male	40	20	60
Female	14	26	40
Total	54	46	100

(c) *Use the information in the table to answer the following questions.*

  (i) *Among the males surveyed, what percentage said they eat at a fast-food restaurant at least once a week?*

  (ii) *Among the females surveyed, what percentage said they eat at a fast-food restaurant at least once a week?*

(d) *Does there appear to be an association between gender and eating at a fast-food restaurant at least once a week? Justify your answer.*

Example 3: Describing the relationship between two quantitative variables by interpreting a least-squares regression line
(**from** LOCUS Project)

The heights (in centimeters) and arm spans (in centimeters) of 31 students were measured. The association between x (height) and y (arm span) is shown in the scatterplot (see Figure 5). The equation of the least-squares regression line for this association is also given.

$$\text{estimated armspan} = 4.5 + 0.977\text{height}$$

(a) If Mike is 5 cm taller than George, what is the expected difference in their arm spans? Show your work.

(b) Jane is 158 cm tall and has an arm span of 154 cm. Rhonda is 163 cm tall and has an arm span of 165 cm. Does the least-squares regression line give a more accurate predicted value for Jane or Rhonda? Explain.

(c) Doug is 210 cm tall. Would you use this least-squares regression line to predict his arm span? Explain.

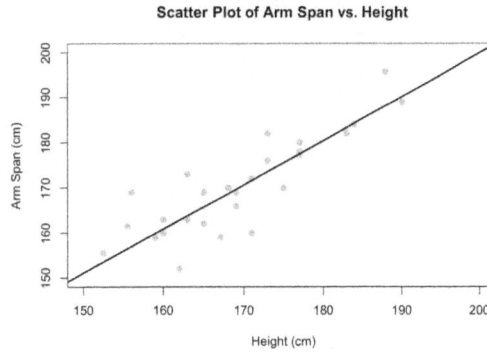

Figure 5: Scatterplot with least-squares regression line

Example 4: Simulation and deciding if an observed statistic is unusual or plausible (**from** LOCUS Project)

Stella saw the following headline in a national newspaper: "30 Percent of High School Students Favor Extended School Day." She wondered if the percentage of students at her school who favor an extended school day was less than 30 percent. To investigate, she selected a random sample of 50 students from the 1,200 students at her school and asked each student in the sample if he or she favors an extended school day.

Only 12 of the students in the sample favored an extended school day. Because the sample percentage is $(12/50)100 = 24\%$, Stella thinks that fewer than 30 percent of the students at her school favor an extended school day. She wonders if it would be surprising to see a sample percentage of 24 or less if the school percentage is really 30.

(a) To see what values of the sample percentage would be expected if the school percentage was 30, she decides to use 1,200 beads to represent the population of 1,200 students. She will use a red bead to represent a student who favors an extended school day and a white bead to represent a student who does not. How many red beads and how many white beads should Stella use?

Stella put all the beads in a box. After mixing the beads, she selected 50 of them and computed the percentage of red beads. She put the 50 beads back in the box and repeated this process 99 more times. Then, she made a dotplot of the 100 sample percentages (see Figure 6).

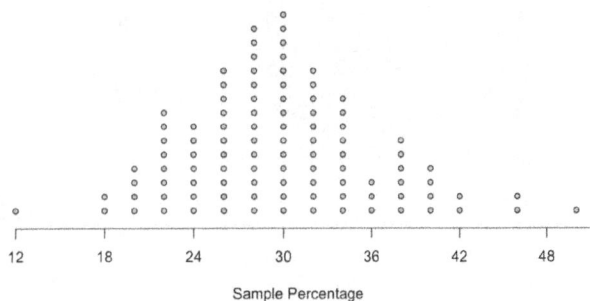

(b) If the school percentage were actually 30%, how surprising would it be to see a sample percentage of 24% or less? Justify your answer using the dotplot.

Figure 6: Dotplot of sample percentages of red beads

(c) Based on her sample data, should Stella conclude that the percentage of students at the school who favor an extended school day is less than 30%? Explain why or why not.

References

American Statistical Association. (n.d.-a). *Census at School—United States*. Census at School. Retrieved January 8, 2020, from https://ww2.amstat.org/censusatschool/

American Statistical Association. (n.d.-b). *Education*. ASA Education. Retrieved January 8, 2020, from https://www.amstat.org/ASA/Education/home.aspx

American Statistical Association. (n.d.). *STatistics Education Web (STEW)*. ASA STEW. Retrieved January 8, 2020, from https://www.amstat.org/ASA/Education/STEW/home.aspx

Angwin, J., Larson, J., Mattu, S., & Kirchner, L. (2016, May 23). *Machine Bias* [Text/html]. ProPublica. https://www.propublica.org/article/machine-bias-risk-assessments-in-criminal-sentencing

Arnold, P., & Franklin, C. (forthcoming). What makes a good statistical question? *Journal of Statistics Education*.

CAUSE. (n.d.). *CAUSEweb | Consortium for the Advancement of Undergraduate Statistics Education*. CAUSE-Consortium for the Advancement of Undergraduate Statistics Education. Retrieved January 8, 2020, from https://www.causeweb.org/cause/

Cobb, G. W., & Moore, D. S. (1997). Mathematics, Statistics, and Teaching. *The American Mathematical Monthly*, *104*(9), 801–823. https://doi.org/10.1080/00029890.1997.11990723

Dominus, S. (2017, October 18). When the revolution came for Amy Cuddy. *The New York Times Magazine*. https://www.nytimes.com/2017/10/18/magazine/when-the-revolution-came-for-amy-cuddy.html

Franklin, C., Bargagliotti, A., Case, C., Kader, G., Scheaffer, R., & Spangler, D. A. (2015). The statistical education of teachers. American Statistical Association. Available at Www.Amstat.Org/Education/SET/SET.Pdf, 124.

Franklin, C., Kader, G., Mewborn, D., Moreno, J., Peck, R., Perry, M., & Scheaffer, R. (2007). *Guidelines for assessment and instruction in statistics education (GAISE) report: A Pre-K–12 curriculum framework*. American Statistical Association.

Gelman, A., & Nolan, D. (2002). You Can Load a Die, But You Can't Bias a Coin. *The American Statistician*, *56*(4), 308–311. https://doi.org/10.1198/000313002605

Hare, L. (n.d.). *Bottle Biology Terrarium*. NSTA Bottle Biology Terrarium. Retrieved January 8, 2020, from https://ngss.nsta.org/Resource.aspx?ResourceID=94

Häusler, N., Haba-Rubio, J., Heinzer, R., & Marques-Vidal, P. (2019). Association of napping with incident cardiovascular events in a prospective cohort study. *Heart*, *105*(23), 1793–1798. https://doi.org/10.1136/heartjnl-2019-314999

Kader, G. D., & Franklin, C. A. (2008). The Evolution of Pearson's Correlation Coefficient. *Mathematics Teacher*, *102*(4), 292–299.

Kuiper, S. (2010, July). Incorporating a research experience into an early undergraduate statistics course. *International Conference on Teaching Statistics (ICOTS)*.

Leavy, A., Meletiou-Mavrotheris, M., & Paparistodemou, E. (2018). *Statistics in Early Childhood and Primary Education: Supporting Early Statistical and Probabilistic Thinking*. https://doi.org/10.1007/978-981-13-1044-7

Levitt, S. (n.d.). *America's Math Curriculum Doesn't Add Up* (No. 391, October 2, 2019). http://freakonomics.com/podcast/math-curriculum/

Lindgren, M. (n.d.). *Detailed income calculations for Dollar Street*. Retrieved January 8, 2020, from https://drive.google.com/drive/folders/0B9jWD65HiLUnRm5ZNWlMSU5GNEU

LOCUS Levels of Conceptual Understanding in Statistics. (n.d.). Retrieved January 9, 2020, from https://locus.statisticseducation.org/

National Council of Teachers of Mathematics (Ed.). (2018). *Catalyzing change in high school mathematics: Initiating critical conversations*. National Council of Teachers of Mathematics.

NGSS Lead States. (2013). *Next Generation Science Standards: For States, By States*. National Academies Press. https://doi.org/10.17226/18290

Nuzzo, R. (2014). Scientific method: Statistical errors. *Nature, 506*(7487), 150–152. https://doi.org/10.1038/506150a

Questionaire Dollar Street. (n.d.). Google Docs. Retrieved January 8, 2020, from https://docs.google.com/document/d/1vtracv6xSEDWvglYVDi7k2fAATs55wZUwnMdECIIfS4/edit?usp=embed_facebook

Rosling Ronnlund, A. (n.d.). *Dollar Street—Photos as data to kill country stereotypes*. Dollar Street. Retrieved January 8, 2020, from https://www.gapminder.org/dollar-street/matrix

Rosling Rönnlund, A. (2017, April). *See how the rest of the world lives, organized by income*. https://www.youtube.com/watch?time_continue=3&v=u4L130DkdOw

Silverman, M. E., Murray, T. J., & Bryan, C. S. (2008). *The quotable Osler* (Rev. pbk. ed). American College of Physicians.

The Learning Network. (2018, January 9). What's Going On in This Graph | January 9, 2018. The New York Times. https://www.nytimes.com/2018/01/04/learning/whats-going-on-in-this-graph-jan-9-2018.html

The Learning Network. (2019, September 25). What's Going On in This Graph | September 25, 2019. The New York Times. https://www.nytimes.com/2019/09/19/learning/whats-going-on-in-this-graph-sept-25-2019.html

Wasserstein, R. L., & Lazar, N. A. (2016). The ASA Statement on p -Values: Context, Process, and Purpose. *The American Statistician, 70*(2), 129–133. https://doi.org/10.1080/00031305.2016.1154108

www.ingramcontent.com/pod-product-compliance
Lightning Source LLC
Chambersburg PA
CBHW051337200326
41519CB00026B/7463